AGRIBUSINESS MANAGEMENT
AN INTRODUCTION

AGRIBUSINESS MANAGEMENT
AN INTRODUCTION

by
Samir Mandal
Aneesha U
Rajesh N

Wordpen Academics
Publishing Science That Matter

ISBN: 978-93-6084-044-0 (Hardback)

Published by : **WORDPEN ACADEMICS**
4736/23, Ansari Road, Darya Ganj,
New Delhi – 110 002
Phone: +91-011-35004336
E-mail: info@wordpenacademics.com

Printed at **:** **Replika Press Pvt. Ltd.**

Exclusive Distributor: ***Bio-Green Books, New Delhi***

PREFACE

Agribusiness Management: An Introduction is designed to provide a foundational understanding of the key principles, functions, and dynamics that shape the world of agribusiness today. As global food systems become increasingly complex and interconnected, the effective management of agricultural enterprises—from small-scale farms to multinational agribusiness firms—has never been more crucial.

This book aims to equip students, early-career professionals, and anyone interested in agriculture with a comprehensive overview of agribusiness operations. It covers essential topics such as production and operations management, marketing, finance, supply chain logistics, and policy frameworks within the agricultural context. Special attention is given to emerging challenges and innovations, including sustainability practices, the role of technology, and the evolving demands of global markets.

We have sought to balance theoretical foundations with practical insights by incorporating real-world examples, case studies, and applied exercises that bridge academic knowledge with industry relevance. The interdisciplinary nature of agribusiness management—drawing on economics, business administration, agricultural sciences, and public policy—is reflected throughout the chapters.

We are grateful to the scholars, industry practitioners, and institutions whose contributions and research have shaped this field and informed the content of this book. Our hope is that this introductory text not only builds competence but also inspires readers to explore the vital role agribusiness plays in feeding the world and supporting sustainable rural development.

Samir Mandal
Aneesha U
Rajesh N

CONTENTS

PREFACE V

1. AN INTRODUCTION 1
   - Agribusiness: Vast and complex system
   - Commerce: major Component of Agribusiness
   - Agribusiness in India
   - Features of Agri-Food Systems
   - Search for Sustainability Solutions
   - Functional Character of an Agribusiness
   - Conclusion
2. AGRICULTURAL ECONOMICS 25
   - Origins
   - Agricultural Extension
   - History
   - Four Paradigms of Agricultural Extension
   - Models of Economic Development
   - Rural Community Development
3. AGRICULTURAL PRODUCTS MARKETING 55
   - Agricultural marketing development
   - Agricultural marketing support
   - Linking farmers to markets
   - Techniques and technologies
   - Terracing
   - Municipal land banks in the United States
   - Agricultural Marketing in India
   - Inadequacies of Present Marketing System
   - Characteristics of Agricultural Product

- Existing systems of agricultural marketing in India
- Improvement of Agricultural Marketing System
- Cooperative Marketing
- Warehousing in India
- Ideal Marketing System
- Commodity Marketing
- Act of Agricultural Marketing

4. CONSUMER DEMAND **113**

- Satisfying consumer demands
- Ensuring food safety
- Changing demands and the agri-food chain
- Agribusinesses in The New Millennium
- Planning for the Future
- Forecasting Methods
- Qualitative Forecasting Methods
- Subjective Approach
- Quantitative Forecasting
- Understanding of Behaviour of Time Series

5. AGRICULTURE INCOME, OUTPUT, PRICE INDICES **131**

- Data sources and availability
- Agricultural income per annual work unit
- Context
- Bakeries dominate manufacturing
- Prices and expenditure
- Definitions of variables and indicators

6. BUDGETING **155**

- Business start-up budget
- Corporate budget
- Organizing for Success
- Information and Technology
- Business Plan for Agribusiness Project
- Transformation of American Agriculture
- Green Revolution
- Strategic Readjustment to the Chinese Economic Structure

- Innovation in Agriculture Technology
- Construction
- Machinery & Equipment

7. ACCOUNTING **207**

- Basic Accounting Documents
- KEEPING farm RECORDS
- Using Accounting Information for Business Control and Planning
- Management Information Systems
- Enterprise Resource Planning

8. HUMAN RESOURCE **223**

- Leadership
- Human Resources Leadership Development Program
- Changing Management Style
- Value of Context of HRM in India
- Managing People on the Farm
- Middle Management in Agriculture
- Middle Management HRM Practices Models
- A business for multitalented people
- Agricultural Economics and Agribusiness
- Conversational skills
- Skills for successful selling
- Sales-force Compensation

BIBLIOGRAPHY **255**

INDEX **257**

1

An Introduction

The food and fiber system is recently being referred to as "agribusiness". The term agribusiness was coined by Davis and Goldberg in 1957 which it represents three part system made up of:

1. the agricultural input sector
2. the production sector and
3. the processing-manufacturing sector. To capture the full meaning of the term "agribusiness", it is important to visualizes these there sectors as interrelated parts of a system in which the success of each part depends heavily on the proper functioning of the other two.

In agriculture, agribusiness is a generat term for the various businesses involved in food production, including farming and, wholesale and distribution, processing, marketing, and retail sales. The term agribusiness has two distinctly different connotations depending on context.

Agribusiness system has undergone a rapid transformation as new industries have evolve and traditional farming operations have grown larger and more specialized. The transformation did not happen over night, but came slowly as a response to a variety of forces. Knowing something about how agribusiness came about makes it easier to understand how this system operates today and how it is likely to change in the future.

Among critics of large-scale, industrialized, vertically integrated food production, the term agribusiness is used negatively, synonymous with corporate farming. As such, it is often contrasted with smaller family-owned farms. Negative connotations are also derived from the negative associations of "business" and "corporations" by critics of capitalism or corporate excess. As concern over global warming intensifies, biofuels derived from food crops quickly emerged as a practical answer to the energy crisis.

Adding corn ethanol to gasoline or using palm oil for biodiese lmakes the fuel burn more cleanly, stretches oil supplies, and perhaps most attractive to some politicians, provides a nice boost to big agribusiness. In Europe and in the US, increasing biofuels was mandated by law. Rising fuel costs are increasingly adding financial burdens on the day-to-day running of agricultural companies.

Input farms are major part of agribusiness and produce variety of technologically based products that account for approximately 75 per cent of all the inputs used in production agriculture. Examples of agribusinesses include Monsanto, seed and agrichemical producer; ADM, grain transport and processing; John Deere, farm machinery producer; Ocean Spray, farmer's cooperative; and Purina Farms, agri-tourism farm.

The word agriculture indicate plowing a field, planting seed, harvesting a crop, milking cows, or feeding livestock. Until recently, this was a fairly accurate picture. But to days' agriculture is radically different.

AGRIBUSINESS: VAST AND COMPLEX SYSTEM

Agriculture has evolved into agribusiness and has become a vast and complex system that reaches for beyond the farm to include all those who are involved in bringing food and fiber to consumers. Agribusiness include not only those that farm the land but also the people and firms that provide the inputs like Seed, chemicals, credit etc., process the output like Milk, grain, meat etc., manufacture the food products like icc-cream, bread, breakfast cereals etc., and transport and sell the food products to consumers as restaurants, supermarkets.

Initially agriculture being the major venture it was easy to become a farmer, but yield was low. Average Indian farmer produced enough food to feed just four people. As a consequence most farmers were nearly totally self-sufficient. They produced most of the inputs they needed for production, such as seed, draft animals, feed and simple farm equipment. Farm families processed the commodities they grew to make their own food and clothing. They consumed or used just about everything they produced.

The small amount of output not consumed on the farm was sold for cash. These items were used to feed and cloth the minor portion of the country's population that lived in villages and cities. A few agricultural products made their way into the export market and were sold to buyers is other countries.

Farmers found it increasingly profitable to concentrate on production and began to purchase inputs they formerly made themselves. This trend enabled others to build business that focused on meeting the need for inputs used in production agriculture such as seed, fencing, machinery and so on.

These farms involved into the industries that make up the "agricultural inputs sector".

At the same time, the agriculture input sector was evolving, a similar evaluation was taking place a commodity processing and food manufacturing moved off the farm. The form of most commodities like wheat, rice, milk, livestock and so on that must be changed to make them more useful and convenient for consumers. For ex. consumers would rather buy flour than grind the wheat themselves before backing a cake. They are willing to pay extra for the convenience of buying the processed commodity flour instead of the raw agriculture commodity like wheat.

During the same period technological advance were being made in food preservation method. Up until this time the perishable nature of most agriculture commodities meant that they were available only at harvest. Advance in food processing have made it possible to get those commodities all throughout the year.

Today, even most farm families use purchased food and fiber products rather than doing the processing themselves. The farms that meet the consumers demand for greater processing and convenience also constitute a major part of agribusiness and are referred to as the processing manufacturing sector.

The evolution from agriculture to agribusiness has brought with it numerous benefits. These include reduced drudgery for laborers; the release of workers for nonagricultural endeavors; a better quality of food and fibers; a greater variety of products; improved nutrition; and increased mobility of people. The release of farm manpower and the creation of new, off-the-farm jobs have been the basis for the country's economic growth and development for the last 150 years. The key to this growth and development has been increased worker productivity, which in turn spurs creativity, new products and wealth. This translates into risk capital, new factories, new jobs, and increased consumer purchasing power.

It is apparent that the definition of agriculture had to be expanded to include more than production. Farmers rely on the input industries to provide the products and service they need to produce agricultural commodities. They also rely on commodity processors, food manufactures, and ultimately food distributors and retailers to purchase their raw agricultural commodities and to process and deliver them to the consumer for final sale. The result is the food and fiber system.

Within the agriculture industry, agribusiness is widely used simply as a convenient portmanteau of agriculture and business, referring to the range of activities and disciplines encompassed by modern food production. There are academic degrees in and departments of agribusiness, agribusiness trade associations, agribusiness publications, and so forth, worldwide. Here, the term is only descriptive, and is synonymous in the broadest sense with food industry. The UN's Food and Agriculture Organization, for example, operates a section devoted to Agribusiness Development , which seeks to promote food industry growth in the Third World.

Changing taste of consumers

Today the business has become very competitive and complex. This is mainly due to changing taste and fashion of the consumers on the one hand, and introduction of substitute and cheaper and better competitive goods, on the other. The old dictum "produce and sells has changed overtime into "produce only what customers want". In fact, knowing what customers want in never simple.

Nevertheless, a farmer operator/farmer manager has to give proper thought to this consideration in order to make his business a successful one. The important requisites for success in a modern business are:

1. Determination of objectives: It is one of the most essential pre requisite for the success of business. The objectives set forth should be realistic and clearly defined. Then, all the business efforts should be geared to achieve the set objectives. In a way, objectives are destination points for an agribusiness. As a car driver must know here he/she has to reach, i.e. destination similarly business also must know what objectives.

2. Proper Planning: In simple words, planning is a pre-determined line of action. The accomplishment of objectives set, to a great extent, depends upon planning itself. It is said that it does not take time to do thing but it takes time to decide what and how to do. Planning is a proposal based on part experience and present trends for future actions. In other words, it is an analysis of a problem and finding out the solutions to solve them with reference to the objective of the farm.

3. Symmetrical organization: An organization is the art or science of building up systematical whole by a number of but related parts. Just as car is build up by various parts like engine, tyre, gear box etc. similarly, organization of business is a harmonies combination of men, machine material, money management etc. so that all these could work jointly as one unit, i.e. "business" "the agribusiness". Organization is, thus such a systematic combination of various related parts for achieving a defined objective in an effective manner.

4. Influence of Consumer Behaviour: As indicated earlier, today the agricultural production philosophy "produce what the consumer want". "Consumers" behaviour is influenced by variety of factors like cultural, social, personal and psychological factors. The business needs to know and appreciate these factors and then function accordingly. The knowledge of these factors is acquired through market research.

Research is a systematic search for new knowledge. Market research enable a business in finding out new methods of production, enhancing the quality of product and developing new products as per the changing tastes and wants if the consumers.

5. Finance: It is said to be the heart & soul of business enterprise. It combined by brings together the land, labour, machine and raw materials into production. Agribusiness should estimate its financial requirements adequately so that it may keep the business wheel on moving. Therefore, proper arrangements should be made for securing the required finance for the enterprise.

6. Right location, right place: The parameter of agribusiness success depends to mainly on the location. Where it is set up. Location of the business should be convenient from various points of view such as availability of required infrastructure facilities, availability of inputs like raw materials, skill labour, nearer to the market etc.

Hence the business men must take sufficient care in the initial stages to selected suitable location for his business.

The sine of the business is also important because the requirement for infrastructural facilities and inputs varies as per the size of the business. The requirement for raw materials, for example, will be les in a smaller sized firm than a larger size firm.

7. Efficient management: One of the reasons for failure of business often attributed to as their poor management or inefficient management. The one man, i.e. the proprietor may not be equally good in all areas of the business. Efficient businessman can make proper use of available resources for achieving the objectives set for the business.

8. Harmonious relations: In an agribusiness organization, the farmer operator occupies a distinct place because he/she is the main living factor among all factors of production.

In fact, it is the human factor who makes the use of other non-human factors like land, machine, money etc. Therefore, for successful operation of business, there should be cordial and harmonious relations maintained with the workers/labours to get their full cooperation in achieving business activities.

Scope of agribusiness

As it has been already mentioned that agribusiness is a complex, system of input sector, production sector, processing manufacturing sector and transport and marketing sector.

Therefore, it is directly related to industry, commence and trade, industry is concerned with the production of commodities and materials while commerce and trade are concerned with their distribution.

Industry

Industry refers to the processes of extraction and production of goods meant for final consumption or use buy individual or buy another industry for its production. Thus, goods used by the final or ultimate consumers are called "consumer goods" such as edible oils, fruit jams, papaya, pickles etc.

Types of Industries

As per the agribusiness concerned, the industries are broadly classified into following types.

(1) **Extractive industries:** These industries are concerned with the extraction; and utilization of natural resources. Example are fishing, fruit gathering, agro-based industries, forestation.

(2) **Genetic Industries:** These industries include breeding of plants, seeds, cattle breeding farm, fish hatcheries, poultry farms. Of course, factors like nature, climate and environment play a dominant role in these industries, yet human skill involved in their production cannot be ignored. For example intensive agriculture is possible with greater amount of capital and larger number of workers.

(3) Manufacturing Industries: These industries are engaged in the conversion of raw material or semi-finished goods produced in the extractive industries. Some prominent examples are – cotton textile industry, spinning and weaving mills etc. Manufacturing industries can further be classified into five types : (i) Analytical industry (ii) Processing industry (iii) Synthetic industry, (iv) Service industry (v) Assembly industry.

COMMERCE: MAJOR COMPONENT OF AGRIBUSINESS

Commerce is the another major component of agribusiness. It includes all those activities which are necessary to bring goods and services from the place of their production to the place of their consumption. Thus, sit includes the buying and selling of goods and service and all those activities which facilitate trade such as storing, grading, packaging, financing, insurance and transportation.

In simple words, commerce includes trade and aid to trade. The principal function of trade (commerce) are to remove the hindrance of person, place, time exchange, knowledge etc. and ensure a free and smooth flow of goods from the producers to the consumers.

Trade in fact is a branch of commerce itself. In a way, it is the final state of business activity involving sale and purchase of commodities or goods. It does not include and to trade like transportation, insurance, banking, finance etc. On the basis of its coverage and volume, trade is normally classified into the following types :

1. On the basis of volume
   - (i) Wholesale trade
   - (ii) Retail trade
2. On the basis of coverage
   - (i) Regional trade
   - (ii) National trade

Agribusiness is a broad concept used to describe corporate agricultural enterprises individually and collectively. Agribusinesses are companies involved in one or more stages of the production of crops and livestock. Examples of agribusiness activities include:

- Manufacture or distribution of agricultural supplies and equipment such as machinery, feed, and fertilizers
- Ownership or management of agricultural production facilities such as farmlands and livestock facilities
- Research and development of new agricultural resources and methods
- Processing or distribution of agricultural products

Providing food or fibers is the ultimate product of all agribusiness operations. As such, the economic impact of agribusiness is significant; agribusiness is almost two times as large as the sum of all manufacturing enterprises (measured in total assets); it represents 40 percent of all consumer spending; and it employs 37 percent of the labor force.

Agribusiness approach

The agribusiness approach is a method of examining farming problems in a new and more comprehensive setting. One benefit from this approach has been the release of workers—farm manpower—from agriculture for employment in new non-farm occupations—including the armed forces during wars. This has resulted in tremendous economic growth and development and an improved standard of living.

In the early nineteenth century, agriculture was a self-contained industry. The typical farm family produced its own food, fuel, shelter, draft animals, feed, tools, implements, and even clothing. Only a few necessities had to be bartered for or purchased off the farm. The farm family performed virtually all operations pertaining to the production, processing, storage, and distribution of farm commodities.

In the ensuing years, however, agriculture evolved from self-sufficiency to intricate interdependence with other segments of the economy, particularly those relating to the manufacture of production supplies and the processing and distribution of food and fiber products.

Agribusiness comparises of several million farm units and several thousand business units, each an independent entity, free to make its own decisions. Agribusiness is the sum total of hundreds of trade associations, commodity organizations, farm organizations, quasi-research bodies, conference bodies, and committees, each concentrating on its own interests. The U.S. government also is a part of agribusiness to the degree that it is involved in research, the regulation of food and fiber operations, and the ownership and trading of farm commodities. Land-grant colleges, with their teaching, experiment stations, and extension functions, form another sector of agribusiness. Agribusiness exists in a vast mosaic of decentralized entities, functions, and operations relating to food and fiber.

Use of New Technologies

One prominent feature of agribusiness is its continuous pursuit of new technologies. Well-known examples include the use of satellite-based global positioning systems (GPS) to closely manage crop lands and computer systems to manage various parts of the business. These technologies boost agricultural efficiency by reducing wasted resources, saving time, and improving output.

Crop agriculture, in particular, has turned to such high-tech solutions to develop what is known as precision or site-specific farming methods. These methods involve systematically testing crop fields for variations in fertility and soil composition. The data are then stored in a computer, and, using GPS equipment on the farm machinery, the on-board computer can then determine where in the fields to allocate seed and fertilizer to maximize yield and minimize waste.

Agribusiness

A number of colleges and universities offer extensive course work in agribusiness and agribusiness degrees. Graduate programs lead to an MBA in agribusiness or a master of science in agribusiness. Typical classes in graduate programs deal with:

- Agribusiness management
- Agricultural industries marketing
- Financial management
- Commodity trading
- Technology

The emphasis of these programs is on developing business acumen and learning about the best practices in the various fields of agribusiness.

AGRIBUSINESS IN INDIA

As incomes in India rise, tastes are changing. Staples are receding in importance. In 1983 food grains accounted for 32% of India's agricultural output; 15 years later they accounted for less than a quarter. But food grains still occupy nearly 64% of India's farmland and an even higher fraction of government concern and energy. In India, cereals are being increasingly overtaken by fruits, vegetables, milk, eggs and poultry.

India is the third-biggest producer of potatoes in the world. The humble spud finds itself stuffed into flatbread, encrusted in cumin seeds or tucked into pancakes. But the truckloads of large, oblong potatoes that arrive at the McCain Foods plant in the Mehsana district of Gujarat face a more exacting ordeal. Ferried by a conveyor belt and propelled by water, they are sized, steam-peeled, sliced, diced, blanched, dried, fried (for precisely 42 seconds in vegetable oil at 199ºC), chilled, frozen, bagged and then boxed.

The fifteen kg boxes of fries that emerge at the other end of this pipeline supply the growing chain of McDonald's restaurants in India. When McDonald's first entered India in 1996, the food-processing industry was confined largely to ice-cream and ketchup. Even importing frozen fries was complicated by the fact that such an exotic item did not appear on India's schedule of tariffs and quotas. It took McDonald's roughly six years and $100m to weld a reliable supply chain together.

For fries, that supply chain begins with 2,000 acres of potato fields in Gujarat, cultivated by 400 farmers under contract with McCain Foods. These cultivators belong to a profession which still employs about half of India's workforce. In the home state of Mahatma Gandhi, agriculture is growing almost as quickly as the rest of the Indian economy. Gandhiji used to say that agriculture is the backbone of Indian economy. But elsewhere, agriculture is said to be in crisis. The average size of farmers' landholdings is only about 1.3 hectares.

If their fields are irrigated at all, they are flooded wastefully, with water flowing down furrows on either side of the crop, taking valuable nitrogen with it. India produces more tractors than any other country, but many farmers still use bullocks instead. They sell their produce at controlled prices in government markets: marketplaces regulated by the state with the aim of protecting farmers from exploitation by unscrupulous traders.

State governments once took it upon themselves to spread know-how, market intelligence and the fruits of agricultural research to smallholders. But the agricultural extension system is now in some disrepair. Public investment in agriculture has stagnated over the past few years as the government's subsidy bill for food, fertiliser and fuel has risen.

In the dearth of public investment, Indian agriculture is increasingly dependent on private outlays, which now account for three-quarters of total investment in the sector. McCain Foods, for example, invested $25m in the Mehsana plant. And in the absence of government extension services, some private companies are finding alternative ways to let farmers know what the customer wants, and how to fulfil that expectation.

Hitesh Patel, for example, used to grow cottonseed on his six hectares in Idar village, about 125km from the Mehsana plant. Some years ago he planted a hectare of potatoes at McCain Foods' urging and under its guidance. Now, he plants potatoes on all of the six hectares he owns and another 1.6 he has leased.

McCain Foods offers him an assured price of 6.50 rupees ($0.14) per kilogram of potatoes, a better rate than the it. But they will not buy just any potato. India's common varieties are too small, watery and sugary, caramelising when fried.

Potatoes fit for processing are usually grown in more temperate latitudes. The tuber likes warm days and cool nights. "Wherever are the best regions for wine, potato is not far behind," says Ghislain Pelletier, a McCain agronomist. Firms before McCain had tried and failed to produce potatoes in India that were suitable for processing into fries. Even Harrison McCain, one of the company's founders, doubted it could be done.

But after several years of experimentation, McCain can now supply all of McDonald's needs in India, as well as producing some creations of its own, like Masala Fries. "We used to struggle for size; now we struggle to reduce the size," says Devendra Kumar of McCain India.

McCain's agronomists would first visit farmers every other day. Now they check in once a week. They insist that farmers give up flooding their furrows with water in favour of drip irrigation, which flows through a pipe punctuated with small nozzles laid along the crop-bed.

Drip irrigation produces "more crop per drop", as Mr Pelletier puts it. It moistens the soil at each root, but leaves the ground otherwise dry. This in turn reduces the humidity that attracts pests and blight.

This kind of contract farming began with Punjab farmers growing tomatoes for Pepsi's food business in the 1980s. But its spread was hampered by tight regulations at the state level on who could buy produce and how. Most states have now eased those restrictions, raising hopes that contract farming—which only accounts for a small fraction of India's annual agricultural output of $220 billion—will flourish.

McDonald's, for example, recently introduced the Chicken McNugget to India, where it counts as a premium product: all white meat without any bone. When the firm first entered the country, it struggled to find poultry suppliers who could debone meat. Because the chain's products are delicate or perishable they must be handled with far more care than rice or wheat. Frozen fries, for example, are so brittle that they must be handled like eggs. But McDonald's also found that the few refrigerated lorries in the country were mostly devoted to transporting ice-cream.

Investments by multinationals like McDonald's can spread into the wider economy. McDonald's, for example, invited East Balt Commissary of Chicago, whose founder supplied buns to Ray Kroc's first McDonald's franchise, to go to India to train the Cremica bakeries in Delhi and Mumbai. East Balt is what Abhijit Upadhye of McDonald's India calls a "good system player".

It also asked Schreiber International, which supplies sliced cheese, to tie up with Dynamix Dairy in India. As a result, many other companies, including Nestlé, Unilever and Pepsi, started doing brisk business with Dynamix.

Moreover, scores of research and development projects are conducted in agribusiness to find new technologies and to better use existing ones. Important R&D work includes developing genetically engineered crops, improving the pest resistance of crops, using biotechnology in agriculture, and formulating new agricultural pharmaceuticals and chemicals. Despite considerable government and university backing for agribusiness research, the majority of this research is funded by the private sector.

The Global Agri-Food System

Technological advancement in twenty first century has given new dimension to agri-food system.

In an era of rapid change and growing risk and uncertainty, agricultural policy and practice in the developing world is encountering, a number of limitations which reveal inadequacies in its long-term sustainability and its capacity to meet the range of objectives that it is expected to deliver. These include concerns about chronic hunger and malnutrition, adverse environmental changes, the limits of technology-enhanced productivity gains, increasing land exploitation and the loss of biodiversity, livelihood insecurity and the continuing poverty of agricultural communities. Worries about food safety, hygiene and nutrition, and growing demands for the re-localisation of agri-food systems from citizen consumers in both the North and the South have also emerged. These apprehensions raise important questions about whether the forms of agriculture developed over the past century, and celebrated as technically advanced and 'modern', are able to respond to the complex and diverse array of in this new era.

FEATURES OF AGRI-FOOD SYSTEMS

Agriculture therefore has a number of features that distinguish it from other productive sectors. Among other things, it is mainly a private activity implemented locally mostly by households, but it also has many dimensions of collective action, is deeply affected by global forces and depends greatly on public interventions for its structure, its support and its development. Taking into account these drivers of change, what are the key characteristics of agri-food systems that demand attention? The following, we suggest, are critical:

- The dynamics of production. Agriculture is characterised by high dependency on natural resources, spatial dispersion of activity, seasonal variables, asymmetries in information due to location and distance, high risks associated with the vagaries of nature, and difficulties in sustaining the productivity of natural resources, because their use and reproduction typically conflict.
- Integrated agri-food systems. Market developments, technological progress, institutional changes and policy interventions in one part of the world have far-reaching implications, even for distant actors, as global and regional supply chains link producers and consumers in different parts of the world.
- Market failures. Failures of input and output markets for agricultural goods and services are associated with high transactions costs, particularly adversely affecting poor farmers, information asymmetries, incomplete property rights, externalities and missing actors. These affect access to markets, the availability of insurance and financial services, and the underwriting of contracts.
- Public sector interventions. There is a signiûcant need for public sector interventions to compensate for these market failures. This makes agriculture highly vulnerable to extractive policies (cheap food policies), land grabbing, rent seeking, regressive subsidies and exposure to corrupt officials. Public budgets in agriculture can easily be prey to clientelism and elite capture, major causes for the mis-investment of already undersized public budgets.

  Equally, they can be distorted and misdirected through the changing whims and misguided policy prescriptions of international donors. Consequently, the political economy of policy and investment in agriculture can determine success or failure in agricultural development.
- Socio-cultural systems. The close correspondence between agriculture as a productive activity ('agri-') and rurality as a way of life ('-culture') make social relations in rural society important determinants of access to resources (for example through land rental markets), asymmetries in power (including by gender and ethnicity) and benefits from public services. These affect dynamic poverty outcomes in agriculture.
- Heterogeneity and diversity. Actors with better asset endowments in favourable areas can take advantage of new markets and of new technological and institutional opportunities. In contrast, large segments of the smallholder population remain reliant on subsistence-oriented activities, linked to labour markets as net sellers and to food markets as net buyers, but relying on agriculture for home consumption and as a safety net of last resort.
- Collective action. Eûective forms of cooperation and collective action are essential for the millions of smallholders, pastoralists, fisher folk and farmworkers to have their voices heard in key regional and national policy forums. They are also essential to enable producers' associations and federations achieve economies of scale (meeting new market requirements (e.g. grades and

standards) and interacting in local clusters of economic activity), access public services and manage common property resources.

The key question – addressed in a preliminary way by this paper and the wider work on food and agriculture by the STEPS Centre – is: how, in the face of these old and new challenges, can poor, marginal people negotiate pathways to sustainability through agriculture? This paper therefore provides an assessment of how different visions of agricultural development, underpinned by contrasting narratives of technological and economic change, respond to these challenges. Yet, despite the diversity of drivers and contexts, the complexity of dynamics and the uncertainties that prevail, current debates about agricultural change in developing countries are often couched – implicitly, if not explicitly – in terms of notions of 'progress' towards a singular goal.

Frequently, the underlying assumption is that such progress is achieved through the transfer of knowledge, ideas, models, practices and technologies from the 'developed' world to the 'developing' world, or from 'modern' science to 'backward' farming settings. For example, in the early 1990s some Indian agricultural universities based their curriculum on two decade old textbooks from the United States, as if no new, locally-speciûc innovations had happened in the intervening years. Within this framing are often embedded notions about how agricultural development occurs in a linear sequence of stages – from 'backward' to 'modern', from 'old' to 'new', from 'underdeveloped' to progressively more 'developed', from pre-industrial to industrial. Thus, there is often assumed to be a singular path to progress, and to be committed to this path governments, farmers, aid agencies and analysis have to be uniformly and unquestioningly 'pro-innovation', 'pro-technology' and 'proudevelopment'.

Those who criticise this monolithic linear assumption are sometimes accused of being 'anti-technology' or 'anti-modernity'. Our approach is not hostile to beneûcial technological changes, but we do reject approaches that assume that there is one, and only one, technological trajectory that implicitly denies the existence and benefits of alternative pathways, or even multiple pathways, towards a broader goal of poverty reduction, social justice and environmental sustainability.

Thus, in considering possible pathways towards this goal, we assume that there may be multiple routes to improving the relationship between complex food system and poor people in developing countries, and that poor rural people often have relevant agricultural knowledge.

Agricultural Sustainability

The term "agribusiness" was coined in the 1950s by John Herbert Davis and Ray A. Goldberg to reflect the two-way interdependence between businesspeople and farmers in the dual roles of suppliers and purchasers. Business firms that serve agriculture rely on farmers for their markets and for some of their supplies. By the same token, farms could not operate without businesses that manufacture farm supplies and those that store, process, and merchandise farm commodities.

Over the past 20 years, a great deal of work on agricultural sustainability has focused on the capacity of food systems to absorb perturbations and still maintain their functions. In a resilient and robust agri-food system, disturbances have the potential to create opportunities for doing new things, for innovation and for new pathways of development. In a vulnerable system, even small disturbances may cause signiûcant adverse social consequences especially for those who are most vulnerable, such as the rural poor in develop ing countries. Until recently, dominant perspectives in conventional agricultural science and development programmes have implicitly assumed a stable and almost indeûnitely resilient environment, where resource flows could be controlled and nature would return to a steady state when human stressors were removed.

Such static, equilibrium-centred views, we argue, provide inadequate insight into the dynamic character of agri-food systems, particularly in an era of global economic and environmental change, where factors such as climate change, rapid land use shifts and uncertain political economic conditions in agricultural economies all impinge on the day-today realities of poorer producers and consumers in the developing world.

Our focus on uncertainty, complexity and diversity is at the core of the STEPS Centre agenda, and aims to shift attention from policies and practices that aspire to maintain the *status quo* or control change in systems assumed to be stable, in favour of analysis and practices that enhance the capacity of agri-food systems to respond to, cope with and shape change. Such responses in turn enhance the possibilities of sustaining desirable, yet diverse, pathways for development in changing environments, where the future is unpredictable and surprises are likely.

Agricultural & Resource Management Problems

Much conventional agricultural science and policy does not seem to be able to explain, let alone respond to, complexity, diversity, uncertainty and non-equilibrium states, although poor people who are dependent on agriculture for their livelihoods very often live in complex, diverse and risk-prone settings, with inherent seasonal instability.

Vulnerability not only damages people's welfare, it also reduces growth, both directly by destroying assets and indirectly as threats of shocks and stresses cause assets to be diverted assets from more productive activities to those that reduce risk and uncertainty.

Agricultural and resource management problems typically tend to be classic 'systems' problems, where aspects of systems behaviour are both complex and unpredictable and where causes, while at times apparently simple, when ûnally understood are always multiple. These problems are often non-linear in nature, cross-scale in time and space and dynamic in character. This is true for both natural and social systems and their interactions. In fact, they need to be understood as one system, with critical feedbacks across temporal and spatial scales. Thus, interdisciplinary and integrated modes of inquiry are needed for understanding and designing eûec-tive responses to human–environment interactions related to food and agriculture in a turbulent world.

A critical minority of policy-makers and citizens – both producers and con-sumers are demanding integrated solutions that address these issues of uncertainty, iversity and complexity. Their calls for action are not so much driven by prophesies of doom as by the need for understanding and action. But if you seek understanding, to whom do you turn for information and advice?

Agricultural science often provides only limited assistance, largely because it includes not only conflicting voices – witness the debate on genetically modified (GM) crops or arguments over food production vs. population growth – but also conflicting modes of inquiry and criteria for establishing the trustworthiness of different lines of argument.

In particular, the philosophies of two streams of agricultural science are often in opposition. The tension between them is now evident in biology. One stream is represented by the paradigm of molecular biology and genetic engineering. This stream of science promises to provide not only health and economic benefits from agricultural biotechnology, but also an uncertain era of changing social values and consequences.

This stream is a science of parts; that is an analysis of speciûc biophysical processes that affect survival, growth and distribution of target variables as if they were independent of each other and could be systematically controlled one at a time. It emerged from a tradition of experimental science, where a narrow enough focus is chosen to pose speciûc questions and empirical hypotheses, collect data and design critical tests for the rejection of falsiûed hypotheses. The goal is to narrow uncertainty to the point where acceptance of an argument among scientific peers is essentially unanimous. Thus, it is conservative and narrowly focused, and it achieves this by being fragmentary and incomplete. It provides individual build-ing blocks of an ediûce, but not the architectural design. This kind of approach to modern agricultural science, a science of the parts, may be suitable for certain types of conventional agricultural development but not for sustainable agriculture – if sustainable agriculture is deûned more broadly to include a range of ecological, economic and social objectives, such as sustained reductions in chronic malnutri-tion, poverty and ecological harm.

By contrast, a holistic stream can be characterised as a science of integration; that is, by inter-disciplinarity and synthesis, by cross-sectoral and cross-scale research and analyses. It is represented, for example, by agroecology, conservation biology, landscape ecology and other systems approaches that include the analysis of (agri-food)-ecosystems, the interactions between multiple coexisting populations and landscapes, and more recently, the study of socio-ecological dynamics at different scales and concerns about global environmental change, such as climate change.

The applied forms of this stream have emerged regionally in new forms of integrated agricultural practice and natural resource and environmental management, where uncertainty and surprises become an essential part of an anticipated set of adaptive responses. They are fundamentally about blending disciplinary perspectives and combining historical, comparative and experimental approaches at scales appropriate to the issues. It is a stream of investigation that is fundamentally concerned with integrative modes of inquiry and multiple sources of evidence.

SEARCH FOR SUSTAINABLILITY SOLUTIONS

Characteristics of Agri-food Systems

If a consideration of dynamic uncertainty needs to be at the core of any search for sustainable solutions to developing world agriculture, what, then, are the factors that drive change and create risks and uncertainties in developing world agriculture today? This section explores this question through an assessment of key drivers of change and their eûects that characterise contemporary agrifood systems.

Agriculture is an important source of livelihoods in developing countries, providing ways of life for billions of people, many of them poor. Of the world's 6.5 billion inhabitants, 5.5 billion live in developing countries, 3 billion in the rural areas of these countries (World Bank 2007).

Of rural inhabitants, an estimated 2.5 billion are involved in agriculture, 1.3 billion are smallholders, while others include farm labourers, migrant workers, herders, fishers, artisans and indigenous peoples who depend on agriculture and natural resources for their livelihoods. More than half are women. The developing world will remain predominantly rural until around 2020 and millions of poor people in those countries will continue to rely on agriculture for their livelihoods for the foreseeable future.

The contribution of agriculture to livelihoods is evident from the fact that 70 percent of the world's poor people, including the poorest of the poor, and 75 percent of the world's malnourished live in rural areas, where most of them are involved in agriculture. The Millennium Development Goal of halving extreme poverty and hunger will not be met without reducing this rural poverty (UN Millennium Project 2005a).

Meeting this food security goal will be a major challenge. Yet rural poverty remains stubbornly high, even with rapid growth in the rest of the economy. Rural-urban income gaps tend to rise as non-agricultural growth accelerates, creating major social tensions as expectations for better lives remain unfulûlled for a majority of the rural people.

Given these trends, chronic hunger and global food security will remain a worldwide concern for the next fifty years and beyond, as the world's population grows from its current 6.5 billion to upwards of ten billion, most of whom will reside in developing countries. Of course, predictions of food security outcomes have been a part of the policy discourse in agriculture at least since the Reverend

Thomas Malthus wrote *An Essay on the Principle of Population* in 1798. Over the past several decades, some neo-Malthusians or 'catastrophists' have expressed concern about the ability of agricultural production to keep pace with global food demands whereas other 'cornucopians' have forecast that technological advances or expansions of cultivated area would boost production suûciently to meet rising demands. Thus far, dire predictions of a global food security catastrophe have proved unfounded, in the sense that ag- gregate food supply has kept pace with population growth, although hundreds of millions remain hungry and malnourished.

Nevertheless, despite the fact that food production per capita has been increasing globally, major distributional inequalities remain, linked primarily to poverty. Global food production has increased by well over 130 percent since the 1960s, yet the fact that almost 78 per cent of countries that report child malnutrition are food-exporting countries dramatically illustrates a 'paradox of plenty'. Moreover, the productivity of major cereals appears to be reaching biological limits in some regions, despite heavy use of agrochemical inputs, and consequently production is now growing more slowly than in recent decades. Widespread and persistent hunger is a fundamental contradiction in today's world when production and productivity in agriculture have grown faster than eûective demand. An estimated 852 million people were undernourished in 2000–02, up 37 million from the period 1997–99. Of this total, more than 95 percent live in developing countries (World Bank 2006). Sub-Saharan Africa, the region with the largest share of undernourished people, is also the place where per capita food production has lagged the most.

This underperformance of the agriculture sector has been exacerbated by ethnic conflict and political instability, declining terms of trade, dwindling investments in agricultural research and infrastructure and increasing water scarcity.

The challenges faced today are, however, substantially different from those encountered by the green revolution producers who achieved sustained gains in agriculture productivity only a few decades ago. Since the 1980s, there has been a substantial decline in public sector support for agriculture and many producers have lost access to key inputs and services. While public sector provision of these services was never very eûcient, it often provided the linkages to markets for poor rural producers.

Today, such links are tenuous and complicated by much greater integration of the global economy. Smallholder producers now compete in global markets that are much more demanding in terms of grades and standards (e.g. quality, traceability and food safety), and more concentrated and vertically integrated than in the past.

A major concern about this concentration is the control exercised by a handful of private corporations over decision-making throughout the agri-food system. In the past, most of the global trading and grain-handling farms were family-held operations which operated in one or two stages of the food system and in a very few commodities. Consequently, risk exposure of many rural households to market forces was very different from that today. Risks were often reduced by the state through government-controlled marketing boards and similar parastatal organisations, which assured a price structure, input and output markets and access to improved technologies and training. Public investments in research and development resulted in higher yielding farm systems.

Furthermore, innovations were encouraged through public subsidies of one kind or another. In much of Asia and Latin America these innovations led many farm households to shift to more productive and higher return farming systems (World Bank 2005).

Today, the system is becoming much more complex, starting with a farm's involvement in (bio) technology, extending through agro-chemical inputs and production, and ending with highly processed food. Increasingly, these farms are developing a variety of different alliances with other players in the system, forming new food system 'clusters'.

As agriculture becomes more concentrated and integrated, these giant clusters are establishing an oligopsony – a market in which a small number of buyers exerts power over a large number of sellers – over large parts of the agri-food chain, enabling them to maximise profit while minimising risk. As a result, the food system has begun to resemble an hourglass.

At the bottom are millions of farmers and farm labourers producing the food and fibre, while at the top are billions of consumers, both rich and poor. At the narrow point in the middle are the dozen or so multinational corporations – the input suppliers, food processors and retailers – earning a profit from every transaction. Typically, goods are exchanged through closed contracts or intra-farm transfers rather than open wholesale markets and even when they are exchanged in wholesale markets, prices may be well below the cost of production due to oversupply. Consequently, the 'cost-price squeeze' falls on the producers, who bear the bulk of the risk and share little, if any, of the rewards.

Because agriculture has a larger tradable component than most sectors, it is profoundly affected by the trade environment and trade policy. Whereas overall trade barriers in industrial countries have declined signiûcantly over the last decade, the remaining barriers are concentrated on agricultural products and labour-intensive manufactures in which developing countries have a comparative advantage.

High levels of farm support, at the level of USD $279 billion (EUR €226 billion) per year in countries belonging to the OECD, depress world prices for several key commodities (especially sugar, cotton, milk, and beef) and deeply undermine agricultural growth in developing countries. Quotas and tariffs remain important instruments for protection, and sanitary and phytosanitary restrictions increasingly perform the same function.

Global and regional economic integration is accompanied by other challenges that further weaken the socioeconomic position of the rural poor. In some parts of the world, especially in Sub-Saharan Africa, rural areas are hard hit by the HIV/AIDS pandemic, which is disrupting the transfer of knowledge, destroying traditional land allocation systems, and dramatically changing the demographic composition of many rural communities.

Global environmental change is increasing pressure on an already fragile natural resource base in complex, risk-prone environments that are the mainstay of rural livelihoods. Rising energy prices are driving massive investments in biofuels, which could increase the volatility of food prices with negative food security implications in some regions (UN-Energy 2007). Finally, conflict conditions, many of which result from or are provoked by poverty, are further eroding the livelihood systems and resilience of poor rural people.

This is not helped by the fact that attention to agriculture in terms of policy commitments and investment levels declined in both international donor and developing country policies and programmes, despite the demonstrated high rates of return and the reductions in poverty that come from such investments. Further progress is curtailed by weaknesses and deficiencies in agricultural science and technology policy regimes that result in institutional arrangements and organisational forms unsuited to development and broad-based diûusion of poverty-reducing innovations. Investments in science and technology have been shown to pay oû most strongly for countries and regions with highly integrated technical and economic systems able to diffuse and apply results of new research.

Constraints faced in mobilising public resources for agricultural development in countries with widespread poverty and undernourishment are illustrated by relating government expenditure on agriculture to the size of the agricultural labour force. In countries where more than 35 percent of the people are undernourished, government expenditure per agricultural worker averages USD $14 or 50 times less than the USD $880 in countries with the lowest rates of undernourishment.

Similar problems also affect agricultural R&D. Corporate R&D agendas understandably focus on potentially profitable sectors, which frequently do not include poor people. Only in the public and charitable sectors have agricultural research and development policies engaged with the needs of the rural poor.

Assumptions about the vulnerability and/or robustness and resilience of gricultural systems remain contested. Many studies predict that world food supplies may not necessarily be adversely affected by moderate climate change, but only by assuming farmers will take adequate steps to adapt to climate change and that in some regions additional CO2 will contribute positively to increased yields. Many developing countries are likely to fare badly, however, as climate change may result in unpredictable growing conditions, including more intense rainfall events between prolonged dry periods, as well as reduced or more variable water resources for irrigation in tropical environments.

Increasing agricultural expansion into marginal lands and forests may in turn put these areas at greater risk of environmental degradation. Such conditions may promote pests and disease on crops and livestock and increase the incidence of vector-borne diseases in humans, as well as increase soil erosion and desertiûcation.

The HIV/AIDS pandemic is another relevant concern for the sustainability of agriculture. More than 28 million people have died since the first case was reported in 1981. In 2005, AIDS killed 2.8 million people, and an estimated 4.1 million became infected, bringing to 38.6 million the number of people living with the virus around the world. Of this total, 24.5 million of these people live in SubSaharan Africa (where in some countries one in three adults are infected) and 8.3 million live in Asia. In addition to its direct health, economic and social impacts, the disease also affects food security and nutrition.

Adult labour is often reduced or removed entirely from affected households, and those households then have less capacity to produce or buy food, as assets are often depleted for medical and/or funeral costs. The agricultural knowledge-base often deteriorates as individuals with farming experience and scientific knowledge succumb to the disease.

Moreover, agri-food systems are changing in many ways as a result of the dynamic interactions of a range of environmental and socioeconomic drivers, including global environmental change, agricultural intensification, concentration of production, vertical integration and coordination, industrialisation, deregulation and economic liberalisation and urbanisation.

The Agribusiness Manager

It was the industrial sector that had gained a lot of importance over the years, but now, we are seeing a turnaround, the agriculture sector is seeing a boom in business, and many people are moving towards it in search of employment. The business is going great and what do all good businesses require, good managers. To know more about agribusiness managers, read the agribusiness manager job description.

Work of Agribusiness Manager

The agribusiness manager's job is similar to any other manager, the difference is that this business is agriculture and they most probably work on farms. They need to have in depth knowledge about farming, cattle and all other agriculture related businesses. They may not be working in a corporate like office, but they have to be just as analytical.

The agribusiness managers have to have knowledge about the latest market conditions. They need to have good contacts with buyers and suppliers. They have to get the best prices for all the products, because the food prices depend upon the agriculture business, and food prices affect inflation.

Planning and Implementation

The managers have to hire workers required on the farm. They have to provide training to them. They have to manage their shifts and decide their wages. The agribusiness is a seasonal business; hence, the manager has to plan accordingly. What crops to plant and in which season all has to be decided by the managers. The manager also has to arrange for the workers' employment in the off-season.

These tasks are all a part of the agribusiness manager job description. If cattle breeding is the main business the manger is looking after, they have to arrange for regular veterinary checkups for the animals. They have harvest the animals for their by products like milk, fur, meat etc. They have to arrange for buyers and get good prices for the goods

They have to make sure the business is making profits, and they have to maintain the accounts for the business. For the details, read the agribusiness manager job description given below.

Duties and responsibilities of an agribusiness manager

- They have to study and implement methods for plantation and harvesting
- They have to oversee all the processes like plantation and harvesting
- They have to arrange for pesticides in case the plant catch pests
- They have make sure the harvest is well stored and arrange for buyers
- If they package the harvest, the manger will oversee the packaging process too
- The manager has to manage the whole agriculture business, from production, to packaging, to sales, the manager is in charge of all
- They have to hire workers for they business and give them the adequate training as agriculture jobs require specialist training
- They also have to arrange for the workers employment in the off season
- The manager should have good contact with vendors that will sell them quality seeds at good prices
- The manger will look after all the accounts of the business and make sure that they are making profits
- They have to report to the owners of the business on the progress made

Qualifications Required

- A bachelor's degree in business administration or agriculture is pre-requisite
- Long experience working in the field of agriculture and related fields is desirable.

Agribusiness Manager Ratings

AAG's market leading research in the Agri Managed Investments Sector enables us to provide Manager Ratings. This data is derived from our alliance with Ernst & Young and their Corporate Governance Rating and AAG's own rating of Track record.

This is the first time that Manager Ratings have been compiled and represents another first for AAG and the agri MIS sector. Planners and investors can use this data to assist them in selecting Managers of agri tax effective products. Corporate Governance covers the following areas of review and rating: financial management, compliance, operational performance and governance. Track record covers the following areas of review and rating: management, past projects, markets, marketing and project financial performance.

How a business is managed and governed, and the results it has achieved over the time and fundamental pillars of Manager performance and an excellent guide to a quality Management team.

"The information available to planners and investors from AAG allows for a rational, commercial manager and hence investment selection", said Marcus Elgin, Managing Director of AAG. "We recommend however, that investors spread risk of management, commodity and geographic region of growing to create a blend of risk and return."

FUNCTIONAL CHARACTER OF AN AGRIBUSINESS

As a confirmed pragmatist, it is often difficult for me to abstract from the functional character of an agribusiness firm and, instead, view it as a social organization. To be sure, all businesses and the managerial control thereof, involve people. The very essence of any successful business rests on human interaction. But in our haste to analytically assess a business, to financially control its operations, and to quantitatively plan its future, we often forget to equate "business management" with the "leadership function" required of all social organizations.

One would normally expect that any institution-bound academician would spend much of his/her time in the simple act of reading. Unfortunately, this is one of many misconceptions which characterize the life of an academician. Not unlike agribusiness managers, academicians find that the normal "press of business" requires a vast diversity of actions, most of which prevent the individual from becoming deeply involved in the pursuit of leisurely reading. Unlike managers, however, academicians are occasionally afforded a brief respite from their routine press of business to read and pursue thoughts or ideas which would otherwise be judged inconsequential to their employer's day-to-day operational tasks.

As I pursue, and enjoy, just such a respite, I elect to read, and do so rather extensively. Not all my reading is focused on a specific subject, but much is directed towards my personal interests in the practice of business management. Acknowledging that agribusiness managers are rarely afforded even a brief respite from their daily tasks, I would like to share with you a condensation of observations expropriated from my current practice of leisurely reading.

We must be reminded of the fact that all organizations, social and business, are destined to failure in the absence of some form of leadership. Someone must be in charge. In its most vestigial form, this is even true in its natural form for the birds and the bees, the former with their pecking order, and the latter with their queens.

In the matter of human affairs, even those who would reject all forms of traditional leadership, find an undeniable need for leaders, themselves. I recently had an opportunity to debate with a Libertarian Party member of the Alaska state legislature.

His views were just short of those advocating complete anarchy in the affairs of state. Yet, ironically, it was this same party whose continued political influence was almost totally dependent upon a finely-tuned and well-disciplined party leadership structure.

The desire to equate leadership with management is an unavoidable one. Like cream, it would seem, leaders naturally rise to the surface where they fulfill managerial roles. Why is it then that managers, unlike cream, do not always represent the best part of the whole? To answer this question, we must recognize that the wizardry of popular leadership has, historically, contributed as much to evil as to good.

Adolph Hitler, for example, proved to be a charismatic leader of some renown, whose ability to stimulate a mass following for his own twisted visions surely propagated human suffering. In this noticeable case, when did leadership end and demagoguery begin? In a recent book titled *Leadership*, James MacGregor he wrote, "A leader and a tyrant are polar opposites." Quite clearly, Burns would distinguish between those who would lead and those who would abuse their position of influence.

Managing Agribusiness

This question was asked some months ago during a week-long agribusiness management training program. A team of five participants were asked to meet together as a team and produce a reasonable answer. It was expected that their answer would be lengthy and rather complex. In fact, they returned the following morning with a brief,but rather profound response. In their view, management required the organization of things, while leadership required the organization of people. In a book titled *The Unconscious Conspiracy*, author Warren Bennis wrote in 1976, "leading does not mean managing." Clearly, our team of management trainees and Mr. Bennis were hinting at a similar distinction.

It would seem that a manager who lacks the leadership quality would function mostly in accordance with previously established routines. Of course, such routines require little thought and even less imagination. They may, in fact, be inappropriate routines or they may be functionally obsolete. Yet managers lacking leadership skills will focus on things rather than people and cling to routine procedures as a defensive means for preventing needed changes.

A reliance on routines may not even be a conscious effort as we are all creatures of habit. Each of us seeks and sustains a sense of security through an adherence to established routines. When such an adherence is allowed to supersede a concern for people and the collective well-being of a business, then management ceases to be an analogue to leadership.

Leadership Role

Even those agribusiness managers who actively pursue a leadership role, find themselves inadvertently submerged in protectionist attempts to preserve the status quo. A noted management scientist named Henry Mintzberg once conducted an in-depth analysis of the working patterns of five top corporate executives. Mintzberg concluded that only rarely did these talented individuals have time to think about anything except the particular issue immediately before them.

Nearly half of the decisions made were rendered within a time period of ten minutes or less. In only ten percent of the cases did these individuals have over one hour to ponder and evaluate a specific matter.

In reality, the executive suite was but a frenzy of activity as a steady stream of correspondence and visits filled the entire working day. Under even the best of intentions to the contrary, these persons were forced to rely heavily on routines as the sole means for dealing with the press of business. If your daily work pattern is anything like that of the executives observed, it would seem as if you are running as fast as you can only to stay where you are. Under such conditions, how can anyone find the time and initiative to function as a leader? This question also arose in the course of the aforementioned management training program.

Perhaps, the answer lies in the manager's willingness and ability to ask himself if the use of routine has provided a subconscious excuse for avoiding the more difficult and demanding tasks. Only yesterday I found myself in the act of committing this transgression.

I approached my department chairman with the argument that my service on a multiplicity of university committees (routinely expected of all academicians) had prevented me from finding the free time needed to complete a project. Who was I kidding? My administrator knew as well as did I that free time is made not found. The argument that I could not find the required free time was but a poor excuse for the fact that I had subconsciously sought not to take the time.

Leadership Pattern

In 1958, Robert Tannenbaum and Warren H. Schmidt published a research paper in the Harvard Business Review titled "How to Choose a Leadership Pattern." In this paper, the authors described a so-called "dominance scale" from which evolved alternative managerial styles. The anagement profession later accepted this dominance scale base for the classification of managerial styles which varied from Autocrat to Humanist. By 1973, however, these same authors felt compelled to produce a sequel to their earlier thoughts.

This sequel was prepared on the basis of perceived changes in social patterns that had taken place since 1958. The fifteen years that had passed since their earlier literary work had witnessed the rapid rise of concerns for civil rights, ecology, consumer protection and the quality of life in the workplace. In their opinion, such societal changes had placed upon managers a burden which previously had hardly existed, i.e., the heightened need for a sense of human sensitivity and flexibility in management.

What this suggests is that as a manager/leader in contemporary times, you are more likely to deal with employers with a higher self-image of their role in the business.

They are less willing to accept their position as a subordinate, more willing to question any source of authority, and more demanding of a position of influence in the business. In fact, a new breed of workers emerged from the 1970s period of social turmoil. While tempers have cooled somewhat and younger people have returned to more pragmatic interests, the modern day managers must recognize that now, more than ever before, leadership becomes a matter of eliciting support and cooperation from employees rather than commanding obedience. Recent studies show that employees are more concerned with a sense of personal autonomy, appreciation of their efforts, and an expanded opportunity to fulfill their individual potentials.

Essential Task of Management

If the manager fails to capitalize on the energies supporting these employee goals, those same energies are sure to become the business's loss. Management scholar Douglas McGregor argues that, "the essential task of management is to arrange organizational conditions and methods of operation so that employees can achieve their own goals best by directing their own efforts towards those objectives set by the organization." The managers of the "old school" might argue that such a permissive environment will only result in diminished productivity.

But allowing for the fact that all managers must, at times, play a dictatorial role, it would seem that employees would respond more positively to such moments of toughness, if it is encompassed within a past record of individual respect, consideration and fair play.

CONCLUSION

In conclusion, a rather simple rule emerges, i.e., the respectful treatment of others is likely to be reciprocated. And it is this reciprocation that transforms a manager into a leader. Lao-Tsi was a Chinese poet and philosopher who lived 2,500 years ago. He likely possessed little experience as a manager, yet his words of advice are still timely and relevant.

He wrote, "Fail to honor people, and they will fail to honor you; but of a good leader, who talks little, when his work is done, his aim fulfilled, they will say: we did this ourselves."

2

Agricultural Economics

Agricultural economics originally applied the principles of economics to the production of crops and livestock — a discipline known as agronomics. Agronomics was a branch of economics that specifically dealt with land usage. It focused on maximizing the yield of crops while maintaining a good soil ecosystem. Throughout the 20th century the discipline expanded and the current scope of the discipline is much broader. Agricultural economics today includes a variety of applied areas, having considerable overlap with conventional economics.

ORIGINS

Economics is the study of resource allocation under scarcity. Agronomics, or the application of economic methods to optimizing the decisions made by agricultural producers, grew to prominence around the turn of the 20th century. The field of agricultural economics can be traced out to works on land economics. Henry Charles Taylor was the greatest contributor with the establishment of the Department of Agricultural Economics at Wisconsin. Another contributor, Theodore Schultz was among the first to examine development economics as a problem related directly to agriculture. Schultz was also instrumental in establishing econometrics as a tool for use in analyzing agricultural economics empirically; he noted in his landmark 1956 article that agricultural supply analysis is rooted in "shifting sand," implying that it was and is simply not being done correctly.

Evolution of Agricultural Economics

The field of agricultural economics has evolved over many decades. One scholar summarizes its development as follows:

"Agricultural economics arose in the late 19th century, combined the theory of the firm with marketing and organization theory, and developed throughout the 20th century largely as an empirical branch of general economics. The discipline was closely linked to empirical applications of mathematical statistics and made early and significant contributions to econometric methods. In the 1960's and afterwards, as agricultural sectors in the OECD countries contracted, agricultural economists were drawn to the development problems of poor countries, to the trade and macroeconomic policy implications of agriculture in rich countries, and to a variety of production, consumption, and environmental and resource problems."

Agricultural economists have made many well-known contributions to the economics field with such models as the cobweb model, hedonic regression pricing models, new technology and diffusion models , multifactor productivity and efficiency theory and measurement, and the random coefficients regression. The farm sector is frequently cited as a prime example of the perfect competition economic paradigm.

Since the 1970s, agricultural economics has primarily focused on seven main topics, according to a scholar in the field: technical change and human capital; agricultural environment and resources; risk and uncertainty; consumption and food supply chains; prices and incomes; market structures; and trade and development.

Areas of Concentration

- Econometrics
- International development
- Community and rural development
- Food safety and nutrition
- International trade
- Natural resource and environmental economics
- Production economics
- Risk and uncertainty
- Consumer behavior and household economics
- Health economics
- Labor economics
- Forestry economics
- Analysis of markets and competition
- Agribusiness
- Agricultural marketing
- Agricultural policy
- Industrial organization
- Marketing of agricultural products
- Rural sociology

Agricultural economics tends to be more microeconomic oriented. Many undergraduate Agricultural Economics degrees given by US land-grant universities tend to be more like a traditional business degree rather than a traditional economics degree. At the graduate level, many agricultural economics programs focus on a wide variety of applied microeconomic topics. During the last decades, graduates from Agricultural Economics departments across America find jobs in diversified sectors of the economy; from corporations to government. Their demand is driven by their pragmatism, optimization and decision making skills, and their skills in statistical modelling.

Agrarian System

An agrarian system is a concept used to describe the dynamic set of economic and technological factors that affect agricultural practices. It is premised on the idea that different systems have developed depending on the natural and social conditions specific to a particular region. Political factors also have a bearing on an agrarian system due to issues such as land ownership, labour organization, and forms of cultivation.

As food security has become more important, mostly due to the explosive population growth during the 20th century, the efficiency of agrarian systems has come under greater review.

Types

The basis for a prevailing agrarian system may be derived from one of a number of major types, including agrarian social structure, for example, tribal or ethnic divisions, feudal classes or family based systems. Farming methods such as migratory herding of livestock are a common framework for which an agrarian system may evolve. Other important kinds of system are based on the dominant polictal ideology such as communism or agrarian socialism.

Regional Examples

The Ottoman agrarian system was based around the tapu, which involved a permanent lease of state-owned arable land to a peasant family. In Haiti there was a social system based on collective labor teams, called kounbit, where farms were run by nuclear families and exchanges. This was replaced by smaller groups, called eskouad, who operated on a reciprocal basis or conducted collective labor to other peasants for a price.

Agribusiness

In agriculture, agribusiness is a generic term that refers to the various businesses involved in food production, including farming and contract farming, seed supply, agrichemicals, farm machinery, wholesale and distribution, processing, marketing, and retail sales. The term has two distinctly different connotations depending on context.

Within the agriculture industry, agribusiness is widely used simply as a convenient portmanteau of agriculture and business, referring to the range of activities and disciplines encompassed by modern food production. There are academic degrees in and departments of agribusiness, agribusiness trade associations, agribusiness publications, and so forth, worldwide. Here, the term is only descriptive, and is synonymous in the broadest sense with food industry.

Among critics of large-scale, industrialized, vertically integrated food production, the term *agribusiness* is used as a negative, synonymous with corporate farming. As such, it is often contrasted with family farm. Some negative connotation is also derived from the negative associations of "business" and "corporation" from critics of capitalism or corporate excess. As concern over global warming intensifies, biofuels derived from food crops quickly emerged as a practical answer to the energy crisis. Adding corn ethanol to gasoline or using palm oil for biodiesel makes the fuel burn more cleanly, stretches oil supplies, and perhaps most attractive to some politicians, provides a nice boost to big agribusiness. In Europe and in the US, increasing biofuels was mandated by law.

An example of an agribusiness was the Old North State Winegrowers Co-operative in North Carolina. Wine grape farmers came together to not only sell their grapes but to share a winery, winemaker and marketing brand together. The cooperative failed in 2006, three years after opening its winery.

Many progressive agribusinesses are now operating online businesses. Rising fuel costs are increasingly adding financial burdens on the day to day running of agricultural companies.

AGRICULTURAL EXTENSION

Agricultural extension was once known as the application of scientific research and new knowledge to agricultural practices through farmer education. The field of extension now encompasses a wider range of communication and learning activities organised for rural people by professionals from different disciplines, including agriculture, agricultural marketing, health, and business studies.

Extension practitioners can be found throughout the world, usually working for government agencies. They are represented by several professional organizations (such as APEN), networks (such as AGREN) and extension journals (such as Journal of Extension).

Agricultural extension agencies in developing countries have received large amounts of support from international development organisations such as the World Bank and the Food and Agriculture Organization(FAO) of the United Nations.

Extension Terminology

The term *extension* was first used to describe adult education programmes in England in the second half of the 19th century; these programmes helped to expand - or extend - the work of universities beyond the campus and into the neighbouring community. The term was later adopted in the United States of America, while in Britain it was replaced with "advisory service" in the 20th century. A number of other terms are used in different parts of the world to describe the same or a similar concept:

- Arabic: Al-Ershad ("Guidance")
- Dutch: Voorlichting ("lighting the path")
- German: Beratung ("advisory work")
- French: Vulgarisation ("simplification")
- Spanish: Capacitacion ("improving skills")
- Thai, Lao: Song-Suem ("to promote")
- Persian: Tarvij & Gostaresh ("to promote and to extend")

In the US, an *extension agent* is a university employee who develops and delivers educational programs to assist people in economic and community development, leadership, family issues, agriculture and environment. Another program area extension agents provide is 4-H and Youth. Many extension agents work for cooperative extension service programs at land-grant universities. They are sometimes referred to as county agents or educators.

Definitions of Extension

There is no widely accepted definition of agricultural extension. The ten examples given below are taken from a number of books on extension published over a period of more than 50 years:

- 1949: The central task of extension is to help rural families help themselves by applying science, whether physical or social, to the daily routines of farming, homemaking, and family and community living.
- 1965: Agricultural extension has been described as a system of out-of-school education for rural people.
- 1966: Extension personnel have the task of bringing scientific knowledge to farm families in the farms and homes. The object of the task is to improve the efficiency of agriculture.
- 1973: Extension is a service or system which assists farm people, through educational procedures, in improving farming methods and techniques, increasing production efficiency and income, bettering their levels of living and lifting social and educational standards.
- 1974: Extension involves the conscious use of communication of information to help people form sound opinions and make good decisions.
- 1982: Agricultural Extension: Assistance to farmers to help them identify and analyse their production problems and become aware of the opportunities for improvement.
- 1988: Extension is a professional communication intervention deployed by an institution to induce change in voluntary behaviours with a presumed public or collective utility.
- 1997: Extension [is] the organized exchange of information and the purposive ransfer of skills.
- 1999: The essence of agricultural extension is to facilitate interplay and nurture synergies within a total information system involving agricultural research, agricultural education and a vast complex of information-providing businesses.
- 2004: Extension [is] a series of embedded communicative interventions that are meant, among others, to develop and/or induce innovations which supposedly help to resolve (usually multi-actor) problematic situations.

HISTORY

Origins of Agricultural Extension

Men and women have been growing crops and raising livestock for approximately 10,000 years. Throughout this period, farmers have continually adapted their technologies, assessed the results, and shared what they have learned with other members of the community. Most of this communication has taken the form of verbal explanations and practical demonstrations, but some information took a more durable form as soon as systems of writing were developed. Details of agricultural practices have been found in records from ancient Egypt, Mesopotamia and China going back more than 3,000 years.

It is not known where or when the first extension activities took place. It is known, however, that Chinese officials were creating agricultural policies, documenting practical knowledge, and disseminating advice to farmers at least 2,000 years ago. For example, in approximately 800 BC, the minister responsible for agriculture under one of the Zhou dynasty emperors organized the teaching of crop rotation and drainage to farmers. The minister also leased equipment to farmers, built grain stores and supplied free food during times of famine.

The birth of the modern extension service has been attributed to events that took place in Ireland in the middle of the 19th century. Between 1845–51 the Irish potato crop was destroyed by fungal diseases and a severe famine occurred. The British Government arranged for "practical instructors" to travel to rural areas and teach small farmer how to cultivate alternative crops. This scheme attracted the attention of government officials in Germany, who organized their own system of traveling instructors. By the end of the 19th century, the idea had spread to Denmark, Netherlands, Italy, and France.

The term "university extension" was first used by the Universities of Cambridge and Oxford in 1867 to describe teaching activities that extended the work of the institution beyond the campus. Most of these early activities were not, however, related to agriculture. It was not until the beginning of the 20th century, when colleges in the United States started conducting demonstrations at agricultural shows and giving lectures to farmer's clubs, that the term "extension service" was applied to the type of work that we now recognize by that name.

In the United States, the Hatch Act of 1887 established a system of agricultural experiment stations in conjunction with each state's land-grant university, and the Smith-Lever Act of 1914 created a system of cooperative extension to be operated by those universities in order to inform people about current developments in agriculture, home economics, and related subjects.

Four Generations of Extension in Asia

The development of extension services in modern Asia has differed from country to country. Despite the variations, it is possible to identify a general sequence of four periods or "generations":

- Colonial agriculture: Experimental stations were established in many Asian countries by the colonial powers. The focus of attention was usually on export crops such as rubber, tea, cotton and sugar. Technical advice was provided to plantation managers and large landowners. Assistance to small farmers who grew subsistence crops was rare, except in times of crisis.
- Diverse top-down extension: After independence, commodity-based extension services emerged from the remnants of the colonial system, with production targets established as part of five-year development plans. In addition, various schemes were initiated to meet the needs of small farmers, with support from foreign donors.
- Unified top-down extension: During the 1970s and '80s, the Training and Visit system (T&V) was introduced by the World Bank. Existing organizations were merged into a single national service. Regular messages were delivered to groups of farmers, promoting the adoption of "green revolution" technologies.
- Diverse bottom-up extension: When World Bank funding came to an end, the T&V system collapsed in many countries, leaving behind a patchwork of programmes and projects funded from various other sources. The decline of central planning, combined with a growing concern for sustainability and equity, has resulted in participatory methods gradually replacing top-down approaches.

The fourth generation is well established in some countries, while it has only just begun in other places. While it seems likely that participatory approaches will continue to spread in the next few years, it is impossible to predict the long-term future of extension. Compared to 20 years ago, agricultural extension now receives considerably less support from donor agencies. Among academics working in this field, some have recently argued that agricultural extension *needs to be reinvented as a professional practice*. Other authors have abandoned the idea of extension as a distinct concept, and prefer to think in terms of "knowledge systems" in which farmers are seen as experts rather than adopters.

Communication Processes within Extension Systems

The term "extension" has been used to cover widely differing communication systems. Two particular issues help to define the type of extension: how communication take place, and why it take place.

How Communication takes Place in an Extension System: Paternalism Versus Participation

Early books on extension often describe a model of communication that involved the transmission of messages from "senders" to "receivers". As part of this model, senders are usually people in authority, such as government planners, researchers, and extension staff, while receivers are usually farmers who are relatively poor and uneducated. Although this model might include something called "feedback", it is clear that the senders are in control of the communication process.

The transmission model of communication is closely related to the idea that extension workers are the link (i.e. message carriers) between researchers (senders) and farmers (receivers). Extension programmes based on this model has been described as "paternalistic"; in other words, the actors in the communication process have a parent/child or teacher/student relationship. Other authors have used the term "top-down" to describe these programmes.

In many countries, paternalistic extension is gradually being replaced by more participatory approaches, in which the knowledge and opinions of farmers is considered to be just as important as that of researchers or government officials. Participatory approaches involve information-sharing and joint decision-making. The terms "interactive" and "bottom-up" have been used to describe these approaches.

The development of participatory extension requires a re-examination of the communication process. At the present time, no single description has replaced the transmission model that is referred to above, but two ideas are becoming widely accepted:

- Communication in the context of participatory extension cannot usefully be described in a linear manner with distinct groups of senders and receivers. Instead, extension activities take place within a *knowledge system* consisting of many actors who play different roles at different times.
- Although some actors in the knowledge system have more authority than others, communication usually involves a *negotiation* rather than a transmission. What takes place is a dialogue, with actors collaborating in the construction of shared meanings rather than simply exchanging information.

Why Communication takes Place: Persuasion Versus Education

Although extension programmes have many different goals, most programmes fall into one of two basic categories:

- Systems of communication that aim to change the behaviour of rural people
- Systems of communication that aim to change the knowledge of rural people

There is, of course, a close relationship between knowledge and behaviour; changes in the former often lead to a change in the latter.

If government policy-makers, project managers or researchers direct the topics addressed and projects undertaken, then the purpose of extension is to change behaviour. This approach to extension has been variously described as *directive extension*, *social marketing* and *propaganda*.

If farmers and other rural people direct the extension towards their own needs, then the purpose of extension is changing knowledge. This knowledge helps rural people make their own decisions regarding farming practices. This approach to extension is closely related to *non-formal education* and *conscientization*.

FOUR PARADIGMS OF AGRICULTURAL EXTENSION

Any particular extension system can be described both in terms of both *how* communication takes place and *why* it takes place. It is not the case that paternalistic systems are always persuasive, nor is it the case that participatory projects are necessarily educational. Instead there are four possible combinations, each of which represents a different extension paradigm, as follows:

- Technology Transfer. This paradigm was prevalent in colonial times, and reappeared in the 1970's and 1980's when the Training and Visit system was established across Asia. Technology transfer involves a top-down approach that delivers specific recommendations to farmers about the practices they should adopt.
- Advisory work. This paradigm can be seen today where government organisations or private consulting companies respond to farmers enquiries with technical prescriptions. It also takes the form of projects managed by donor agencies and NGOs that use participatory approaches to promote pre-determined packages of technology.
- Human Resource Development . This paradigm dominated the earliest days of extension in Europe and North America, when universities gave training to rural people who were too poor to attend full-time courses. It continues today in the outreach activities of colleges around the world. Top-down teaching methods are employed, but students are expected to make their own decisions about how to use the knowledge they acquire.
- · Facilitation for empowerment . This paradigm involves methods such as experiential learning and farmer-to-farmer exchanges. Knowledge is gained through interactive processes and the participants are encouraged to make their own decisions. The best know examples in Asia are projects that use Farmer Field Schools (FFS) or participatory technology development (PTD).

It must be noted that there is some disagreement about whether or not the concept and name of *extension* really encompasses all four paradigms. Some experts believe that the term should be restricted to persuasive approaches, while others believe it should only be used for educational activities. Paulo Freire has argued that the terms 'extension' and 'participation' are contradictory. There are philosophical reasons behind these disagreements. From a practical point of view, however, communication processes that conform to each of these four paradigms are currently being organized under the name of extension in one part of the world or another. Pragmatically, if not ideologically, all of these activities *are* agricultural extension.

Extension Methods

Compact area group approach (CAGA) is an innovative extension approach developed by Dr.K. Abdul Kareem, an agricultural extensionist of India to combat problems like coconut eryophyid mite in states like Kerala

Creative Extension is a new branch of agricultural extension pioneered by the Krishi Vigyan Kendra, Kannur under Kerala Agricultural University. Creative Extension uses art form for communication of innovation to farmers and calls for inputs from people with histrionic talents.

Custom Harvesting

In agriculture, custom harvesting or custom combining is the business of harvesting of crops for others. Custom harvesters usually own their own combines and work for the same farms every harvest season. Custom harvesting relieves farmers from having to invest capital in expensive equipment while at the same time maximizing the machinery's use.

The custom harvesting industry has its roots in the mid-twentieth century. Before the invention of the combine harvester, farmers usually owned their own harvesting machinery and worked in tandem with migrant workers, who would bring their own threshing equipment. As combines became more and more widespread, the demand for migrant labor decreased. Custom harvesters, owning their own combines, existed beginning in the 1920s, albeit on a small scale. However, World War II caused a labor and materials shortage, and the custom harvesting industry experienced a great growth. In North America, the area from Texas to Saskatchewan was heavily traversed by custom harvesters. In 1942, there were approximately 500 such operators in this range, but by 1947 that number had grown to 8,000 in Kansas alone. Over the years, the industry transformed from one being run solely by men into family-run businesses. Harvesters would pass their routes and clients down to their children, thus turning custom combining into family tradition.

Custom combiners charge for their services in several ways: a flat per-unit of land price, a per-bale price, or a percentage of the crop. One of the biggest problems for farmers who hire custom harvesters is the availability of labor and the timing of the harvest. If the harvester cannot be available right when the crops are ready, there is the risk of low harvest (if too early) or spoilage (if too late).

The number of custom harvesters has dropped in modern times due to a number of reasons. For one, the consolidation of farms into large-scale operations has allowed them to operate on economies of scale such that they do not need to rely on contractors. Competition is another reason. In China, custom harvesting became such a popular profession in the late 20th century that too many people entered the industry and prices became too low to generate a profit.

Economic Development

Economic development refers to increases in the standard of living of a nation's population associated with sustained growth from a simple, low-income economy to a modern, high-income economy. Its scope includes the process and policies by which a nation improves the economic, political, and social well-being of its people. Economic development may also refer to the field comprising policies and efforts that seek to improve the economic well-being and quality of life for a community or region by creating and/or retaining jobs and supporting or growing incomes and the tax base.

Overview

The *University of Iowa's Center for International Finance and Development* states that:

"'Economic development' or 'development' is a term that economists, politicians, and others have used frequently in the 20th century. The concept, however, has been in existence in the West for centuries. Modernization, Westernization, and especially Industrialization are other terms people have used when discussing economic development. Although no one is sure when the concept originated, most people agree that development is closely bound up with the evolution of capitalism and the demise of feudalism."

Among other things, the contemporary social scientific study of economic development encompasses broad theories of the causes of industrial-economic modernization plus organizational and related aspects of enterprise development in modern societies. It embraces sociological-type research relating to business organization and enterprise development from a historical and comparative perspective; specific processes of the evolution (growth, modernization) of markets and management-employee relations; and culturally related cross-national similarities and differences in patterns of industrial organization in contemporary Western societies. On the subject of the nature and causes of the considerable variations that exist in levels of industrial-economic growth and performance internationally, it seeks answers to such questions as: "Why are levels of direct foreign investment and labour productivity significantly higher in some countries than in others?"

Economic Growth vs Economic Development

Economic development refers to social and technological progress. Economic growth is often assumed to indicate the level of economic development. The term "economic growth" refers to the increase (or growth) of a specific measure such as real national income, gross domestic product, or per capita income. National income or product is commonly expressed in terms of a measure of the aggregate value-added output of the domestic economy called gross domestic product (GDP). When the GDP of a nation rises economists refer to it as economic growth.

The term economic development on the other hand, implies much more. It typically refers to improvements in a variety of indicators such as literacy rates, life expectancy, and poverty rates. GDP is a specific measure of economic welfare that does not take into account important aspects such as leisure time, environmental quality, freedom, or social justice. Economic growth of any specific measure is not a sufficient definition of economic development.

MODELS OF ECONOMIC DEVELOPMENT

The 3 building blocks of most growth models are: (1) the production function, (2) the saving function, and (3) the labor supply function (related to population growth). Together with a saving function, growth rate equals s/â (s is the saving rate, and â is the capital-output ratio). Assuming that the capital-output ratio is fixed by technology and does not change in the short run, growth rate is solely determined by the saving rate on the basis of whatever is saved will be invested.

Harrod-Domar Model

The Harrod-Domar Model delineates a functional economic relationship in which the growth rate of gross domestic product (g) depends positively on the national saving ratio (s) and inversely on the national capital/output ratio (k) so that it is written as $g = s / k$. The equation takes its name from a synthesis of analysis of growth by the British economist Sir Roy F. Harrod and the Polish-American economist Evsey Domar. The Harrod-Domar model in the early postwar times was commonly used by developing countries in economic planning. With a target growth rate, and information on the capital output ratio, the required saving rate can be calculated.

Exogenous Growth Model

The exogenous growth model (or neoclassical growth model) of Robert Solow and others places emphasis on the role of technological change. Unlike the Harrod-Domar model, the saving rate will only determine the level of income but not the rate of growth. The sources-of-growth measurement obtained from this model highlights the relative importance of capital accumulation (as in the Harrod-Domar model) and technological change (as in the Neoclassical model) in economic growth. The original Solow (1957) study showed that technological change accounted for almost 90 percent of U.S., economic growth in the late 19th and early 20th centuries. Empirical studies on developing countries have shown different results.

Krugman (1994), who maintained that economic growth in East Asia was based on perspiration (use of more inputs) and not on inspiration (innovations).

Even so, in our postindustrial economy, economic development, including in emerging countries is now more and more based on innovation and knowledge. Creating business clusters is one of the strategies used. One well known example is Bangalore in India, where the software industry has been encouraged by government support including Software Technology Parks.

Information-Led Development

Information-Led Development (ILD) most commonly refers to a development strategy whereby a developing country makes as a primary economic policy focus the creation and development of a national information technology (IT) sector with the express aim of relying on this sector as an engine of growth. Notable examples of such countries are India and the Philippines.

More recently, a new formulation of ILD has emerged. With origins in community economic development in the United States, the new ILD model describes the use of data to generate actionable information or information solutions to development challenges. Examples of this include the inclusion of non-financial payment obligations in consumer credit files, also known as alternative data, and the use of this information in underwriting, as a means to reduce financial exclusion in the United States, where an estimated 54 million Americans are shut out of mainstream credit access as there is insufficient information about them in their credit files to be scored by a credit scoring model. This variant of ILD was pioneered by PERC, a non-profit policy research organization and development intermediary headquartered in Chapel Hill, North Carolina . Other US-based organizations, including Social Compact and the Local Initiatives Support Corporation, employ variants of ILD, but none has applied this internationally except for PERC.

This development model is gaining traction in emerging markets such as Colombia and South Africa, where the data is being used to reduce financial exclusion and facilitate credit access as a means to build wealth and form assets. It is also attracting increasing attention from development agencies, including USAID, the International Finance Corporation, the World Bank Group, and the Consultative Group to Assist the Poor.

Development by GDP

North America, even though one of the slowest growing continents, has stable growth. Most of the faster growing economies are in the Caribbean.

South America has a Boom and Bust growth with high followed by recession growth, most notable in Brazil, however growth has been stabilizing and the whole continent is growing.

Africa has seen the fastest growing but also the slowest growing/declining. From the oil fields which made Angola the 3rd fastest growing country in the world, to Zimbabwe the slowest growing and declining country in the world. Oil in Africa has created 'wealth spots' were a few countries have exceeded their neighbors in wealth. Out of the 10 fastest growing countries in the world, 3 were African. Some countries have in the past been the fastest growing in the world. Equatorial Guinea reached 75% growth in 2004 because of oil reserves.

Europe has one of the most stable growth rates. After the fall of the Soviet Union, there was a period of economic decline in Eastern Europe over the 1990s, followed by recovery in the 2000s. The region is now experiencing growth, particularly in those countries that have recently joined the European Union. If the Caucasus were included, Europe would be one of the fastest growing continents in the world. Most countries are growing at a medium speed; however, many smaller countries exceed 7% and grow exceptionally faster than their neighbors. Out of the 10 fastest growing countries in the world, only 1 is in Europe.

Overall in the 20th century Asia was seen as the area with most growth; however, in the 21st century, most of this has been dominated by China. Some spots of growth are starting to appear in East and even South Asia. Most nations with high populations have seen high growth especially. Out of the 10 fastest growing countries 3 were directly in Asia, and 3 indirectly or partially.

Meanwhile Oceania has seen moderate growth. The only exceptional growth in Oceania has been on Vanuatu.

Some countries have negative growth, most often due to ongoing wars or hyperinflation. These countries include Palestinean territories, Zimbabwe, Fiji and Chad.

Regional Policy

In its broadest sense, policies of economic development encompass three major areas:

- Governments undertaking to meet broad economic objectives such as price stability, high employment, and sustainable growth. Such efforts include monetary and fiscal policies, regulation of financial institutions, trade, and tax policies.
- Programs that provide infrastructure and services such as highways, parks, affordable housing, crime prevention, and K-12 education.
- Job creation and retention through specific efforts in business finance, marketing, neighborhood development, small business development, business retention and expansion, technology transfer, and real estate development. This third category is a primary focus of economic development professionals.

Economic Developers

Economic development, which is thus essentially economics on a social level, has evolved into a professional industry of highly specialized practitioners. The practitioners have two key roles: one is to provide leadership in policy-making, and the other is to administer policy, programs, and projects. Economic development practitioners generally work in public offices on the state, regional, or municipal level, or in public-private partnerships organizations that may be partially funded by local, regional, state, or federal tax money. These economic development organizations (EDOs) function as individual entities and in some cases as departments of local governments. Their role is to seek out new economic opportunities and retain their existing business wealth.

There are numerous other organizations whose primary function is not economic development work in partnership with economic developers. They include the news media, foundations, utilities, schools, health care providers, faith-based organizations, and colleges, universities, and other education or research institutions.

With more than 20,000 professional economic developers employed world wide in this highly specialized industry, the International Economic Development Council [IEDC] headquartered in Washington, D.C. is a non-profit organization dedicated to helping economic developers do their job more effectively and raising the profile of the profession. With over 4,500 members across the US and internationally, serving exclusively the economic development community. Membership represents the entire range of the profession ranging from regional, state, local, rural, urban, and international economic development organizations, as well as chambers of commerce, technology development agencies, utility companies, educational institutions, consultants and redevelopment authorities. Many individual states also have associations comprising economic development professionals and they work closely with IEDC.

There is intense competition between communities, states, and nations for new economic development projects in today's globalized world, and the struggle to attract and retain business is further intensified by the use of many variations of economic incentives to the potential business such as; tax incentives, help with investment capital, donated land and many others. IEDC places significant attention on the various activities undertaken by economic development organizations to help them compete and sustain vibrant communities.

Additionally, the use of community profiling tools and database templates to measure community assets versus other communities is also an important aspect of economic development. Job creation, economic output, and increase in taxable basis are the most common measurement tools. When considering measurement, too much emphasis has been placed on economic developers for "not creating jobs." However, the reality is that economic developers do not typically create jobs, but facilitate the process for existing businesses and start-ups to do so. Therefore, the economic developer must make sure that there are sufficient economic development programs in place to assist the businesses achieve their goals. Those types of programs are usually policy-created and can be local, regional, statewide and national in nature.

RURAL COMMUNITY DEVELOPMENT

Rural community development encompasses a range of approaches and activities that aim to improve the welfare and livelihoods of people living in rural areas. As a branch of community development, these approaches pay attention to social issues particularly community organizing. This is in contrast to other forms of rural development that focus on public works (e.g. rural roads and electrification) and technology (e.g. tools and techniques for improving agricultural production).

Rural community development is important in developing countries where a large part of the population is engaged in farming. Consequently, a range of community development methods have been created and used by organisations involved in international development. Most of these efforts to promote rural community development are led by 'experts' from outside the community such as government officials, staff of Non-governmental organizations and foreign advisers. This has led to a long debate about the issue of participation, in which questions have been raised about the sustainability of these efforts and the extent to which rural people are-or are not–being empowered to make decisions for themselves.

Rural Community Development in the United Kingdom

In the UK rural community development is seen as very important. Rural areas are often some of the most deprived in the country. Rural Community Councils around the country support local rural communities in securing sustainable futures. The local rural communities are supported by experienced community development workers.

Rural Community Development in the United States

In the United States, rural community development is an essential tool in keeping rural areas economically viable in a very competitive global arena. Under the United States Department of Agriculture, this is addressed through the, Rural Development mission area comprised of the Rural Housing Service, Rural Utilities Service, and Rural Business-Cooperative Service. Research and data sources for rural areas of the United States is also addressed by the United States Department of Agriculture through the Economic Research Service and the National Agricultural Library's Rural Information Center.

Part of the United States Department of Commerce, the Economic Development Administration (EDA) is tasked within it's mission "to generate jobs, help retain existing jobs, and stimulate industrial and commercial growth in economically distressed areas of the United States. EDA assistance is available to rural and urban areas of the Nation experiencing high unemployment, low income, or other severe economic distress."

There are four Regional Rural Development Centers in the United States that coordinate "rural development research and extension (education) programs cooperatively with the land-grant institutions regionally and nationally. The Centers support and strengthen individual state efforts in rural areas by developing networks of university research and extension faculty from a variety of disciplines to address rural issues."

Economics in Indian Agriculture

Indian is an agricultural country, where 70 per cent population is dependent on agriculture. This forms the main source of income. The contribution of agriculture in the national income in India is more, hence, it is said that agriculture in India is a backbone of Indian Economy.

The Indian economy is basically agrarian. In spite of economic development and industrialization, agriculture is the backbone of the Indian economy. As

Mahatma Gandhi said, "India lives in villages and agriculture is the soul of Indian economy". Nearly two-thirds of its population depends directly on agriculture for its livelihood. Agriculture is the main stay of India's economy. It contributes about 26 percent of the gross domestic product. Agriculture meets food requirements of the people and produces several raw materials for industries.

From agricultural point of view, India is a unique country. It has vast expanse of level land, rich soils, wild climatic variations suited for various types of crops, ample sunshine and a long growing season. The net sown area in India today is about 143 million hectares. India has the highest percentage of land under cultivation in the world. In spite of the fact that large areas in India, after independence, have been brought under irrigation, only one-third of the cropped area is actually irrigated. The productivity of agriculture is very low. Farming depends mainly upon monsoon rain. Most of the production comprises food crops. About one-third of the land holdings are small, less than one hectare in size. Farmers own their own small prices of land and grow crops primarily for consumption. Even storage facilities for crops are inadequate. Now use of pesticides and fertilizers has increased and large areas have been brought under high yielding variety of seeds. This led to green revolution in several parts of India. This has helped in increasing yields per hectare as well as total production of different crops.

There are many reasons responsible for the low productivity of agriculture. About one-third of land holdings are very small less than one hectare in size. Due to small size of land holdings we cannot use modern way of cultivation. Even today the farmers are using very old methods, tools and implements for fanning. Farmers are not using artificial ways of cultivation. Inputs like-better quality of seeds, fertilizers and pesticides are also not used by most of the farmers. Exploitation of marginal farmers is also responsible. There is also low productivity because of increasing pressure on land and absence of bank credit.

Agriculture is the backbone of our Indian economy. Agricultural development is a precondition of our national prosperity. It is the main source of earning livelihood of the people. Nearly two-thirds of its population depends directly on agriculture. Agriculture provides direct employment to 70 percent of working people in the country. It is the main stay of India's economy.

Apart from those who are directly involved in the agrarian sector, a large number of the population is also engaged in agro-based activities. Agriculture meets the foods requirements of large population of India. It ensures food security for the country. Substantial increase in the production of food grain like-rice, wheat etc. and non-food grains like-tea, coffee, spices, fruits and vegetables, sugar, cotton etc. has made India self-sufficient. Agriculture also contributes to the national income of our country. It accounts for 26 percent of the gross domestic product. The growth of most of the industries depends on agriculture. It produces several materials for industries. It forms the basis of many industries of India like-cotton, textile, jute, sugar industries etc by providing cotton, sugarcane, oilseeds etc. People engaged in agriculture also buy the products of industries like-tractors, pesticides, fertilizers, pump-set etc. Agriculture contributes in foreign exchange of our country. India exports agricultural products like tea, coffee, sugar, tobacco, spices etc and earns foreign currency. Exports from the agricultural sector have helped India in earning valuable foreign exchange and thereby boosting economic development. From above mentioned facts it is very clear that in spite of industrial development still agriculture is the backbone of the Indian economy.

The Five-year Plans accorded priority to the agricultural sector. In the past 50 years the food grain production in the country increased substantially from 51 million tonnes in 1950-51 to 209 million tonnes in 1999-2000. Inspite of the constant rise of population, we have been able to built a food stock of 44.7 million tonnes in 2001. This is because of the technological and institutional reforms in our country. The Indian government took several steps to improve the agricultural condition in the country. The government has encouraged consolidation of land holdings to promote use of modern farm machines. Land reforms were introduced. The government took lands of big land owners away and redistributed to landless labourers. The government abolished the Zamindari System. Modern methods of cultivation were introduced in the country. The government provided better infrastructure facilities such as- irrigation, electricity and transportation. Agricultural equipments such as- tractors, pump harvesters, fertilizers, pesticides were made available to farmers. Getting finance from banks was made easier for the farmers. The crop insurance was another step to protect the farmers against losses caused by crop failure on account of natural calamities like drought, flood cyclone etc. High-yielding varieties of seeds, fertilizers and irrigation gave birth to Green Revolution. All these led to tremendous increase in the production of crops.

The country on an average, has enough in stock to meet the food requirements of its citizens. India has emerged as the largest producer of coconut, ginger, cashew nut, black-peeper and as the second largest producers of fruits and vegetables. The productivity of the land has increased through the years, but has not reached international productivity levels. Indian agriculture has diversified into various sectors and contributes significantly to the nation's economy. But this situation is not likely to remain so easy in the years to come. The population of India is likely to be around 1300 million. This would require a huge amount of food grains along with non-food grains. India has to use its vast potential of agriculture in a systematic and planned manner. We have to develops some of the techniques which the developed countries have been using.

Agriculture is the main sector of Indian economy which is amply powered by the following points:-

1. Share in National Income: The contribution from agriculture has been continuously falling from 55.1% in 1950-51 to 37.6% in 1981-82 & further to 18.5% in 2006-07. But agriculture still continues to be the main sector because it provides livelihood to a majority of the people.

2. Largest Employment Providing Sector: in 1951, 69.5% of the working population was engaged in agriculture. This percentage fell to 66.9% in 1991 & to 56.7% in 2001. However, with rapid increase in population the absolute number of people engaged in agriculture has become exceedingly large.

3. Provision of Food Surplus to the Expanding Population: Because of the heavy pressure of population in labor-surplus economies like India & its rapid increase the demand for food increases at a fast rate. Therefore, unless agriculture is able to continuously increase its surplus of food-grains, a crisis is likely to emerge. Experts foresee that by the end of 11th five year plan (i.e., 2011-2012), the demand for food-grains is expected to increase to 280.6 million tons. Meeting this demand would require 2% growth per annum. The challenge facing the country is clear as during the last 10 years the food-grains have been growing at a meager 0.48%.

4. Contribution to Capital formation: There is a general agreement on the importance of Capital Formation in economic development. Unless the rate of Capital Formation increases to a sufficient high degree, economic development cannot be achieved. Agriculture can play a big role in pushing the Capital Formation in India. Rural sector can transfer labor & capital to the industrial sector which can be effectively used to increase the productivity in the latter.

5. Providing Raw Material to industries: Agriculture provides raw materials to various industries of national importance. Sugar industry, Jute industry, Cotton textile industry, Vanaspati industry are examples of some such industries which depend on agriculture for their development.

6. Market for Industrial Products: Since more than two-thirds of the population of India lives in rural areas, increased rural purchasing power is a valuable stimulus to industrial development.

7. Importance in International Trade: Agriculture constitutes about 75% of the total exports of the country. Such is the importance of agriculture as far as earnings of foreign exchange are concerned.

Agriculture is the backbone of the Indian economy. Despite major emphasis on industrial development during the last four decades, agriculture continues to occupy a place of pride in our economy. The importance of agriculture can be brought out from the following facts:

Share in national income: Although the share of agriculture in the total national income has been gradually decreasing on account of the development of the secondary and tertiary sectors, it still contributed about 18 % of nation income in 2006-07.(in 1950-51, it was 59%)

Source of employment: In India, agriculture is the main source of employment. Even in 2004-05, more than 56% of the total labour force of India is engaged in agriculture and depend on it for their livelihood (1950-51: 69.5%). It becomes evident from this fact that other sectors of the economy could not generate enough employment for the growing population.

Provision of foodgrains: In a developing country like India where a very large proportion of income is spent on food and the population is increasing rapidly, the demand for foodgrains has been increasing at a fast rate. Agriculture In India has played an important role in meeting almost the entire food needs of the people. The production of foodgrains in India has increased from 51 million tones in 1950-51 to 208.3 million tones in 2005-06, i.e. by a little more than 4 times since 1950-51. This has enabled the country to overcome the problems of foodgrain shortages. The country is almost self-sufficient in foodgrains and it no longer depends on import of foodgrains.

Supply of raw materials to industrial sector: Agriculture plays an important role in industrial development. Many industries like cotton industry, jute industries, sugar industries, food processing industries, etc. depends on agriculture for their raw material requirements. Moreover, workers engaged in various industries depend on agriculture for their food requirement.

Market for industrial product: Agriculture provides markets for a large number of industrial products. Since about two thirds of India lives in rural areas, there is a large rural purchasing power which has created a large demand for all types of industrial products. Green revolution has considerably increased the purchasing power of the large farmers substantially in the recent years. Thus for the demand for various products like soaps, detergents, clothes, cycles, scooters, radios, television, torches, lead batteries, etc. has witnessed a marked increase. Likewise, the demand for a variety of agricultural inputs like chemical fertilizer, tractors, pumpsets, pesticides etc. has increased sharply. This has stimulated the development of industries producing these inputs.

Earner of foreign exchange: Agriculture plays an important role in Indian economy as an earner of foreign exchange through exports of agricultural commodities like tea, cotton, coffee, jute, fruits, vegetables, spices, tobacco, sugar, oil, cashew kernels, etc. in the past, export of agricultural products accounted for about 70% of the export earnings of the country. However, with economic development and consequent diversification of our exports, the share of agriculture in total exports has come down to about 10% in 2005-06. All these exports bring valuable foreign exchange to pay for the increased imports of machinery and raw materials required in the non- agricultural sector.

Significance for trade and transport: Agriculture helps in the development of tertiary (or service) sector. For example various means of transport like roadways and railways get bulk of their business from the movement of agricultural commodities and raw materials. A significant part of internal trade constitutes mainly of agricultural products.

Source of revenue for the government: Through the direct contribution of agricultural taxes to the central and state governments is not significant, they get a significant part of their total revenue in terms of land revenue, irrigation charges, taxes imposed on the commodities purchased by the cultivators etc. central government also earns revenue from export duties on agricultural production. Freight charges imposed by Indian Railways for carrying agricultural product generate huge revenue to the central exchequer.

On over all view, India has always been benefited by agriculture. Though the future of India is industrialization, the contribution of agriculture would always prove to be vital for making India a powerful and stable economy in the future.

The importance of agriculture in the National Economy is explained by the following points.

(1) Share of Agriculture in the national income:

Year	Contribution of agriculture in National Income
1950-51	57%
1970-71	43%
1930-31	40%
1938-39	33%

It is seen that agriculture contributes a major share in the national income of India.

Secondly, the share of agriculture in national income has been decreasing. At present the contribution is about 28%. This steady and gradual decline is due to the rapid Industrialization in the difference part of the country.

The contribution of agriculture in national income in case of some development countries are as follows. USA - 3% Canada 4% Australia 5%

This indicates that the role of agriculture in the national income in the developed countries is negligible. More developed a country, smaller is the contribution of agriculture in the national income and vice-versa.

(2) Agriculture as a source of livelihood: In India about 65 to 70 per cent population dependent on agriculture, the population remains almost constant for number of years. On the other hand in the developed countries less than 10 per population dependent on agriculture

(3) Indian agriculture and pattern of employment:

Year	Percentage of working population in agriculture
1961	69.5%
1981	59.4%
While in U.K.	23%
France	7%

(4) Importance of agriculture in Industrial development: Many industries dependent on agriculture, raw material from agriculture is supplied to many industries e.g. sugar industries, Cotton Industries, Paper Industries, tobacco industries, Chilies, turmeric etc. Many industries supply the inputs to the agricultural industry e.g. fertilizers, insecticides, pesticides, implements and machineries like tractors etc.

(5) Role of agriculture in the field of international trade: Many agricultural products like tea, sugar, oilseeds, tobacco, spices contribute the major share in export. In addition to this, we are exporting fruits some vegetables and flowers to the other countries. Now days we are exporting basmati rice to foreign countries. The proportion of agri. goods is to the tune of 50%. In addition to this goods manufactured from agriculture products contribute 20 percent. Thus, agriculture contributes 70% in export.

In addition to the above the role played by agriculture in Indian is as under.

- Many agriculture produce like food grains, fruits are transported by roadways and railways. Thus, it helps in employment of many people in this field.
- If the agricultural production is good, cultivators will earn more income. They will be in position to purchase manufactured products and other inputs required in agriculture. In short, we can say that the prosperity of the country will depend upon the prosperity of agriculture.

Meaning of Consumption and their Classification

By consumption we mean the satisfaction of our want by the use of commodities and services. When we use a commodity, we really use its want satisfying quality of utility. For example when we take a glass of water to quench our thirst we are said to consume water. When we are sitting on chair in the office, we are consuming the chairs. A person is sick, he calls in a Doctor, and he has consumed the Doctor's services. Thus, taking a glass of water means it is actual consumption of water but in case of chair and Doctor's Services, consumption refer to utilization of chair of Doctor's services, consumption refer to utilization of chair of Doctor's services.

Therefore consumption can be defined as.

1. To make use of any commodity of service for the satisfaction of our wants is called consumption.
2. Consumption means destruction of utility.

Types of consumption

Consumption is known as Direct of Final consumption when the goods satisfy human wants directly and immediately. E.g. taking of meals, use of furniture. On the other hand, consumption is called indirect or productive consumption when the goods are not meant for final consumption but for producing other goods which will satisfy human wants, e.g. use of fertilizer in agriculture.

Importance of Consumption

1. The consumption is very importance as it is the beginning as well as the end of all economic activities. A man feels desire and then the makes an effort to satisfy it when the effort has been made the result is the satisfaction of the want. Thus want is the beginning and satisfaction the end of our economic activity.

2. The production, as an economic activity depends upon consumption. Producer is taking production of a certain good because it is going to be consumed by the consumer. Therefore consumption determines the production.
3. The intensity of consumption of particular goods determines the price in a market. Thus consumption also influences the exchange activity. The existence of exchange in an economy is due to consumption only.
4. Distribution, i.e. sharing of incomes to landlords, capitalist labour and organizer is also influenced by consumption. Consumption indicates standard of living and standard of living determines the efficiency. Efficiency determines the share in the national income.

Human Wants and their Classification

Man is a bundle of desires. His wants are infinite in variety and number. Some wants are natural, for example foods, air, clothing and shelter without which existence of man's life is not possible. Similarly wants vary from individual to individual and they multiply with civilization.

Characteristics of Human Wants

1. Human wants are unlimited: Man's mind is so made that he never completely satisfied and hence there is no end to human wants. One want is satisfied another want will crop up to take its place and thus it is never ending cycle of want.
2. Any particular want is satiable: Though the wants are unlimited, but it is possible to satisfy a particular want, provided has the means (resource).
3. Wants are complementary: It is a common experience that we want things in groups. A single article out of group can not satisfy human wants by it self. It needs other things to complete its use e.g. a motor-car needs petrol and mobile oil it starts working. Thus the relationship between motor-car and petrol is complementary.
4. Wants are competitive: Some wants competes to other. We all have a limited amount of money at our disposal; therefore we must choose some things and reject the other. E.g. sugar and jaggery, tea and coffee.
5. Some Wants are both complimentary and competitive: When use of machinery is done the use of labour needs to be reduced. This indicates competitive nature. But to run the machinery the labour is also required and as such it indicates complimentary relationship.
6. Wants are alternative: There are several ways of satisfying a particular want. If we feel thirsty, we can have water, lassi, in summer while coffee, tea in winter. The final choice depends upon availability of money and the relative prices.
7. Wants vary with time place and person: Wants are not always the same. It varies with individual to individual. People want different things at different times and in different places.

8. Wants vary in Urgency and Intensity: All wants are not equally urgent and in tense. Some wants are urgent while some are less urgent.
9. Wants multiply with civilization: With the advancement the wants multiply. Therefore the wants of people living in urban area are more than the villagers. With civilization the demand for radio, T.V, motor-car etc, are increasing.
10. Wants are recur: Some wants are recurring in nature, e.g. food we require again and again.
11. Wants change into habits: If a particular want is regularly satisfied a person becomes used to it and it grows into habit e.g. smoking of cigarette and use of drugs.
12. Wants are influenced by income, salesmanship and advertisement: It income is higher more wants can be satisfied. Many things we buy of particular brands due to salesmanship or advertisement.
13. Wants are the result of custom or convention: As a part of custom and convention we buy many thins. Really they are not required but unlikely we have to purchase it e.g. expenses on social ceremonies.
14. Present wants are more important then future wants: Future is uncertain and hence man is more concerned with the satisfaction of the present wants rather than future wants.

Classification of Wants

The wants can be classified as under.

A. Necessaries: These can be sub divided as:

- Necessaries of existence: The things without which we can not exist e.g. water, food, clothing, shelter.
- Conventional necessaries: The things which we are forced to use by social custom.

B. Comforts: After satisfying our necessaries we desire to have some comforts. For example table and chair for a student help to increase the efficiency. But cushioned costly chair is not a comfort.

C. Luxuries: Luxury means superfluous consumption. After getting comforts, man desire luxury. The luxury articles need not required e.g. gold and silver, costly furniture, etc.

Standard of Living

Standard of Living refers to the necessaries, comforts and luxuries which a person is accustomed to enjoy.

It can be defined as the mode of living. In other words, the standard of living of the people means the quantity and quality of their consumption.

Factors determining standard of living

The factors affecting the standard of living have been discussed below, in respect of country and individual separately.

A. Factors of which the standard of living in a country depends.

1. **Level of national income (output):** The level of national income depends upon the total volume of production in the country. The countries having higher national income enjoy higher standard of living.
2. **Level of productivity:** The national income depends upon the productivity of a person engaged in agriculture, industry, industry or any economic activity. Thus high productivity resulted in high national income and high standard of living. The advanced countries enjoy high standard of living because their productivity is high.
3. **Size of population:** The per capita income can be estimated by the total national income and size of population. Thus if size of population is larger, the per capita income will be smaller and as such standard of living will be lower.
4. **Distribution of National Income:** It the distribution of income is not equal, the standard of living will affect. National income which has been ill distributed resulted into wide disparity. Thought the per capita income is higher, Due to ill distribution of income few rich person only enjoy higher standard of living and the masses of people have to live with extremely low standard.
5. **Level of Education:** Generally it is observed that educated people tend to have high standard of living on the contrary the illiterate people are reluctant to improve the living standard even though they are provided large income.
6. **General Price level:** Different countries are having different price levels. The country having low price level can provide good standard of living to her people and vice-a versa.
7. **Terms of trade:** The terms of trade can be measured by the taking the ratio of price level of its exports to the price level of its imports, thus only physical production is not sufficient to have higher standard of living, the terms of trade is also equally important in this regard.

B. Factors determining individual standard of living:

Following are the important factors on which the standard of living of a person depends

1. Income 2. Size of family 3. Family tradition 4. Education 5. Social customer 6. General price level.

Importance of the law of DMU

1. Basic of economic law and concepts: This law of DMU forms the basis of law of demand, law of Equimarginal utility, elasticity of demand etc.

2. Public finance: The Govt. can impose and justify progressive income tax on the ground of this law, as the income increases, the MU of income diminishes.

3. Businessmen: A businessman or producer can increase the sale of his product by fixing a lower price. Since consumers tend to buy more to equate MU with price, a producer can expect a rise in sale.

Utility Analysis - Law of Equimarginal Utility (EMU)

This law of Equimarginal Utility is another fundamental principle of Economics. It is also known as law of substitution or law of Maximum satisfaction. We have already seen that human wants are unlimited whereas the means to satisfy these wants are strictly limited. It therefore becomes necessary to pick up the most urgent wants that can be satisfied with the money that a consumer has.

In order to get maximum satisfaction out of the funds (money) we have, we carefully weigh the satisfaction obtained from each rupee that we spend. If we find that a rupee spend in one direction has greater utility than in another, we shall go non spending money, on the former (first) commodity, till the satisfaction derived from the last rupee spent in the two cases is equal. In other words, we substitute some units of commodity of greater utility for some units of the commodity of less utility. The results of this substitution will be the MU of the former will fall and that of the latter will rise, till the two marginal utilities are equalized. That is why this law is called the laws of substitution or equimarginal utility.

This law has been illustrated with the help of table given below

Units	Marginal Utility of oranges	Marginal Utility of apples
1	10	8
2	8	6
3	6	4
4	4	2
5	2	0
6	0	2
7	2	4
8	4	6

Suppose apples and oranges are the commodities to be purchased suppose we have go seven rupees to spend. Let us spend three rupees on oranges and four rupees on apples. The utility of 3rd unit of oranges is 6 and that of the 4th unit of apples is 2. As the MU of oranges is higher, we should buy more of oranges and less of apples. Let us substitute one orange for one apple so that we buy four oranges and three apples. Now the MU of both oranges and apples is the same i.e. 4. This arrangement yields maximum satisfaction. Thus total utility of 4 oranges would be 10+8+6+4=23 and of three apples 8+6+4=18 which gives a total utility of 46. The satisfaction given by 4 oranges and 3 apples of on one rupee each is grater than could be obtained by any other the total utility fiends less than 46. Thus, it can be concluded that we obtain maximum satisfaction when we equalize marginal utilities by substituting some unit of the more useful for the less useful commodity.

Limitation of the Law of Equi- Marginal Utility

1. **Ignorance:** If a consumer is ignorant and blindly follows custom, he will may not make wrong use of money.
2. **Inefficient organizer:** The inefficient business organizer will final to achieve the best result from the land, labour and capital. That he employs.
3. **Unlimited resources:** When the resources are sample this law will be meaning less.
4. **Hold of custom and fashion:** It the purchase is strongly influence by customer and fashion he will not obtain maximum satisfaction.
5. Frequent changes in prices of different goods and services are occurred the observance of this law is difficult.

Practical Importance of Law of EMU

1. **Consumption:** A wise consumer acts on this law while arranging his expenditure and obtains maximum satisfaction.
2. **Production:** To obtain maximum net profit, he must substitute one factor of producing to another so as to have most economical combination.
3. **Exchange:** Exchange implies substitution of one thing to another and hence this law is important.
4. **Distribution:** It is on the principle of the marginal productivity that the share of each factor of production is determined.
5. **Public finance:** The Govt. is also guided by this law in public expenditure. The Govt. can expend its revenue (money) in such a way that it will secure maximum welfare of the people.

Consumer's Surplus

Consumer's surplus is one of the most important concepts in Economics. It was expounded by Alfred Marshall. We often find that the price we pay for a commodity is usually less than the satisfaction we derived from its consumption for example, when we purchase a packet of salt, match-box, news paper etc. and consume, the satisfaction derived from those is greater as compared to the price paid for them. This what consumer surplus mean. The concepts can be defined as under.

It is assumed in the above table the price of oranges in the market is Rs. 2.00 per orange. A consume will purchase as may oranges as make his, marginal utility equal to the price. Thus he will purchase 5 orange and pay for each Rs. 2.00. In this way he will spend Rs. 10.00 But the total utility of the 5 oranges is equal to Rs. 30.00. He thus gets a consumer's surplus equal to (30-10) Rs. 20.00

The consumer's surplus can also be found from fourth column of the table. The utility of the first unit of oranges to the consumer is equal to Rs. 10.00, therefore be would be prepared to pay Rs. 10.00 for it's rather than go with out is. But be pays for the first orange only Rs. 2.00, because the price of an orange in the market is Rs. 2.00. Therefore, from the first unit, the consumer is surplus equal to (10-2) = Rs. 8.00, which is written in the fourth column. Similarly the utility of second orange is equal to 8 while the consumer pays Rs. 2.00 for its and therefore obtains (8-2) = Rs. 6.00 as consumer's surplus. From 5th orange the consumer derives satisfaction equal to Rs. 2.00 and as such the consumer's surplus from fifth unit is equal to (2-2) = 0. Thus if we calculate the total utility obtained (i.e. 30) and total amount paid (Rs. 10.00), the consumer's surplus as given in column no four is equal to. Rs. 20.00

Practically however the measurement of consumer's surplus is not simple. There are numerous difficulties to measure consumer's surplus exactly in the market but it is possible to have rough estimate which is of very great practical value.

Criticism: The concept of consumer's surplus has been critised on several grounds.

1. It is said that this concept is imaginary idea. It is very difficult to say how much one is prepared to pay and if it is said it will unreal.
2. It is very difficult to measure exactly. Because different people are prepared to pay different amount (price) and hence it is very difficult to measure total consumer's surplus in the market.
3. This concept does not apply to necessaries. For example, if we ask how much a man be prepared to pay for a glass of water when he is dying of thirst, it is very difficult to say an exact amount. Thus, consumer's surplus in such cases is immeasurable.

Importance of consumer's surplus: This concept is useful in a number of ways.

In public finance: It is very useful to Finance Minister in imposing taxes and fixing the rates. He will impose more taxes on commodities in which consumer's surplus is more.

- To the businessman and monopolistic as they can increase the price of the commodities in which there is large consumer's surplus.
- Comparing advantages of different places.
- Measuring Benefits from international trade.

Production, Factors affecting the production and Factors of Production

Production, in Economics is one of the important activities whatever human being is received goods as a natural gift can not be consumed as such. It requires some processing and then and then only it is consumed. Through processing we transform some goods and services in to another one for example sugarcane into sugar, Cotton into cloth etc. In economics, sugarcane or cotton are termed as inputs factor or raw material while sugar or cloth are termed as output or finished product. Thus the term can be defined as under.

1. Production means transformation of inputs (goods and sieves) into output.
2. Production of wealth or value.
3. Production means creation (addition) of wealth or value.

It may consist not only goods but also services.

Factors affecting the production: Following factors affect production.

1. Natural factors: like climatic conditions, soil type affect production. Production can be diminished due to natural calamities like flood, drought etc.
2. Technical progress: Can positively influence production. Use of improved variety, fertilizers, insecticides etc. can give us more production.
3. Political factors: also affect production positively or negatively. Decisions pertaining to taxation, investment or fiscal. Policies of Govt. influence production.
4. Infrastructure facilities: Like transport, credit, storage etc. are also equally important to have more production.
5. Character of people: determines productivity. The hard workers and sincere workers always produce more and hence it is very important factor which influences production.

3

Agricultural Products Marketing

Agricultural marketing is inferred to cover the services involved in moving an agricultural product from the farm to the consumer. It is also the planning, organizing, directing and handling of agricultural produce in such a way as to satisfy the farmer, producer and the consumer. Numerous interconnected activities are involved in doing this, such as planning production, growing and harvesting, grading, packing and packaging, transport, storage, agro- and food processing, distribution, advertisingand sale. Effectively, the term encompasses the entire range of supply chain operations. However, its key function is to help direct these services, by providing competent and able market information, thereby linking the other operations into an integrated service with targeted outcomes.

AGRICULTURAL MARKETING DEVELOPMENT

Efforts to develop agricultural marketing have, particularly in developing countries, tended to concentrate on a number of areas, specifically infrastructure development; information provision; training of farmers and traders in marketing and post-harvest issues; and support to the development of an appropriate policy environment. In the past, efforts were made to develop government-run marketing bodies but these have tended to become less prominent over the years.

Agricultural Market infrastructure

Efficient marketing infrastructure such as wholesale, retail and assembly markets and storage facilities is essential for cost-effective marketing, to minimize post-harvest losses and to reduce health risks. Markets play an important role in rural development, income generation, food security, and developing rural-market linkages. Experience shows that planners need to be aware of how to design markets that meet a community's social and economic needs and how to choose a suitable site for a new market. In many cases sites are chosen that are inappropriate and result in under-use or even no use of the infrastructure constructed. It is also not sufficient just to build a market: attention needs to be paid to how that market will be managed, operated and maintained.

Rural assembly markets are located in production areas and primarily serve as places where farmers can meet with traders to sell their products. These may be occasional (perhaps weekly) markets, such as haat bazaars in India and Nepal, or permanent. Terminal wholesale markets are located in major metropolitan areas, where produce is finally channelled to consumers through trade between wholesalers and retailers, caterers, etc. The characteristics of wholesale markets have changed considerably as retailing changes in response to urban growth, the increasing role of supermarkets and increased consumer spending capacity. These changes may require responses in the way in which traditional wholesale markets are organized and managed.

Retail marketing systems in western countries have broadly evolved from traditional street markets through to the modern hypermarket or out-of-town shopping center. In developing countries, there remains scope to improve agricultural marketing by constructing new retail markets, despite the growth of supermarkets, although municipalities often view markets primarily as sources of revenue rather than infrastructure requiring development. Effective regulation of markets is essential. Inside a market, both hygiene rules and revenue collection activities have to be enforced. Of equal importance, however, is the maintenance of order outside the market. Licensed traders in a market will not be willing to cooperate in raising standards if they face competition from unlicensed operators outside who do not pay any of the costs involved in providing a proper service.

Market information

Efficient market information can be shown to have positive benefits for farmers and traders. Up-to-date information on prices and other market factors enables farmers to negotiate with traders and also facilitates spatial distribution of products from rural areas to towns and between markets. Most governments in developing countries have tried to provide market information services to farmers, but these have tended to experience problems of sustainability. Moreover, even when they function, the service provided is often insufficient to allow commercial decisions to be made because of time lags between data collection and dissemination. Modern communications technologies open up the possibility for market information services to improve information delivery through SMS on cell phones and the rapid growth of FM radio stations in many developing countries offers the possibility of more localised information services.

In the longer run, the internet may become an effective way of delivering information to farmers. However, problems associated with the cost and accuracy of data collection still remain to be addressed. Even when they have access to market information, farmers often require assistance in interpreting that information. For example, the market price quoted on the radio may refer to a wholesale selling price and farmers may have difficulty in translating this into a realistic price at their local assembly market. Various attempts have been made in developing countries to introduce commercial market information services but these have largely been targeted at traders, commercial farmers or exporters. It is not easy to see how small, poor farmers can generate sufficient income for a commercial service to be profitable although in India a new service introduced by Thomson Reuters was reportedly used by over 100,000 farmers in its first year of operation. Esoko in West Africa attempts to subsidize the cost of such services to farmers by charging access to a more advanced feature set of mobile-based tools to businesses.

Marketing training

Farmers frequently consider marketing as being their major problem. However, while they are able to identify such problems as poor prices, lack of transport and high post-harvest losses, they are often poorly equipped to identify potential solutions. Successful marketing requires learning new skills, new techniques and new ways of obtaining information. Extension officers working with ministries of agriculture or NGOs are often well-trained in agricultural production techniques but usually lack knowledge of marketing or post-harvest handling.

Enabling environments

Agricultural marketing needs to be conducted within a supportive policy, legal, institutional, macro-economic, infrastructural and bureaucratic environment. Traders and others are generally reluctant to make investments in an uncertain policy climate, such as those that restrict imports and exports or internal produce movement. Businesses have difficulty functioning when their trading activities are hampered by excessive bureaucracy. Inappropriate law can distort and reduce the efficiency of the market, increase the costs of doing business and retard the development of a competitive private sector. Poor support institutions, such as agricultural extension services, municipalities that operate markets inefficiently and inadequate export promotion bodies, can be particularly damaging. Poor roads increase the cost of doing business, reduce payments to farmers and increase prices to consumers. Finally, corruption can increase the transaction costs faced by those in the marketing chain.

AGRICULTURAL MARKETING SUPPORT

Most governments have at some stage made efforts to promote agricultural marketing improvements. In the United States the Agricultural Marketing Service (AMS) is a division of USDA and has programs that provide testing, support standardization and grading and offer market news services. AMS oversees marketing agreements and orders research and promotion programs. It also purchases commodities for federal food programs. USDA also provides support to agricultural marketing work at various universities. In the United Kingdom, support for marketing of some commodities was provided before and after the Second World War by boards such as the Milk Marketing Board and the Egg Marketing Board. These boards were closed down in the 1970s. As a colonial power, Britain established marketing boards in many countries, particularly in Africa. Some continue to exist although many were closed at the time of the introduction of structural adjustment measures in the 1990s.

Several developing countries have established government-sponsored marketing or agribusiness units. South Africa, for example, started the National

Agricultural Marketing Council (NAMC) as a response to the deregulation of the agriculture industry and closure of marketing boards in the country. India has the long-established National Institute of Agricultural Marketing. These are primarily research and policy organizations, but other agencies provide facilitating services for marketing channels, such as the provision of infrastructure, market information and documentation support. Examples from the Caribbean include the National Agricultural Marketing Development Corporation (NAMDEVCO) in Trinidad and Tobago and the New Guyana Marketing Corporation in Guyana.

Recent developments

New marketing linkages between agribusiness, large retailers and farmers are gradually being developed, e.g. through contract farming, group marketing and other forms of collective action. Donors and NGOs are paying increasing attention to ways of promoting direct linkages between farmers and buyers within a value chain context. More attention is now being paid to the development of regional markets (e.g. East Africa) and to structured trading systems that should facilitate such developments. The growth of supermarkets, particularly in Latin America and East and South East Asia, is having a significant impact on marketing channels for horticultural, dairy and livestock products.Nevertheless, “spot” markets will continue to be important for many years, necessitating attention to infrastructure improvement such as for retail and wholesale markets.

Energy crop

An energy crop is a plant grown as a low-cost and low-maintenance harvest used to make biofuels, such as bioethanol, or combusted for its energy content to generate electricity or heat. Energy crops are generally categorized as woody or herbaceous plants; many of the latter are grasses of the family Graminaceae.

Commercial energy crops are typically densely planted, high-yielding crop species which are processed to bio-fuel and burnt to generate power. Woody crops such as willow or poplar are widely utilised, as well as temperate grasses such as Miscanthus and Pennisetum purpureum (both known as elephant grass). If carbohydrate content is desired for the production of biogas, whole-crops such as maize, Sudan grass, millet, white sweet clover, and many others can be made into silage and then converted into biogas.

Through genetic modification and application of biotechnology plants can be manipulated to create greater yields, high energy yields can also be realized with existing cultivars.. However, some additional advantages such as reduced associated costs (i.e. costs during the manufacturing process) and less water use can only be accomplished by using Genetically_modified_crops#Biofuel genetically modified crops.

Types

Energy generated by burning plants grown for the purpose, often after the dry matter is pelletized. Energy crops are used for firing power plants, either alone or co-fired with other fuels. Alternatively they may be used for heat or combined heat and power (CHP) production.

To cover the increasing requirements of woody biomass, short rotation coppice (SRC) were applied to agricultural sites. Within this cropping systems fast growing tree species like willows and poplars are planted in growing cycles of three to five years. The cultivation of this cultures is dependent on wet soil conditions and could be an alternative for moist field sieds. However, an influence on local water conditions could not be excluded. This indicates that an establishment should exclude the vicinity to vulnerable wetland ecosystems.

Gas biomass (methane)

Anaerobic digesters or biogas plants can be directly supplemented with energy crops once they have been ensiled into silage. The fastest growing sector of German biofarming has been in the area of "Renewable Energy Crops" on nearly 500,000 ha (1,200,000 acres) of land (2006). Energy crops can also be grown to boost gas yields where feedstocks have a low energy content, such as manures and spoiled grain. It is estimated that the energy yield presently of bioenergy crops converted via silage to methane is about 2 GWh/km (1.8×10 BTU/sq mi). Small mixed cropping enterprises with animals can use a portion of their acreage to grow and convert energy crops and sustain the entire farms energy requirements with about one fifth of the acreage. In Europe and especially Germany, however, this rapid growth has occurred only with substantial government support, as in the German bonus system for renewable energy. Similar developments of integrating crop farming and bioenergy production via silage-methane have been almost entirely overlooked in N. America, where political and structural issues and a huge continued push to centralize energy production has overshadowed positive developments.

Liquid biomass

European production of biodiesel from energy crops has grown steadily in the last decade, principally focused on rapeseed used for oil and energy. Production of oil/biodiesel from rape covers more than 12,000 km² in Germany alone, and has doubled in the past 15 years.Typical yield of oil as pure biodiesel may be is 100,000 L/km (68,000 US gal/sq mi; 57,000 imp gal/sq mi) or more, making biodiesel crops economically attractive, provided sustainable crop rotations exist that are nutrient-balanced and preventative of the spread of disease such as clubroot. Biodiesel yield of soybeans is significantly lower than that of rape.

Bioethanol

Energy crops for biobutanol are grasses. Two leading non-food crops for the production of cellulosic bioethanol are switchgrass and giant miscanthus. There has been a preoccupation with cellulosic bioethanol in America as the agricultural structure supporting biomethane is absent in many regions, with no credits or bonus system in place. Consequently, a lot of private money and investor hopes are being pinned on marketable and patentable innovations in enzyme hydrolysis and the like.

Bioethanol also refers to the technology of using principally corn (maize seed) to make ethanol directly through fermentation, a process that under certain field and process conditions can consume as much energy as is the energy value of the ethanol it produces, therefore being non-sustainable. New developments in converting grain stillage (referred to as distillers grain stillage or DGS) into biogas energy looks promising as a means to improve the poor energy ratio of this type of bioethanol process.

By dedication

Dedicated energy crops are non-food energy crops as giant miscanthus, switchgrass, jatropha, fungi, and algae. Dedicated energy crops are promising cellulose sources that can be sustainably produced in many regions of the United States. Additionally, the green waste byproducts of food and non-food energy crops can be used to produce various biofuels.

Agricultural value chain

The agricultural value chain concept has been used since the beginning of the millennium, primarily by those working in agricultural development in developing countries. Although there is no universally accepted definition of the term, it normally refers to the whole range of goods and services necessary for an agricultural product to move from the farm to the final customer or consumer.

The term value chain was first popularized in a book published in 1985 by Michael Porter, who used it to illustrate how companies could achieve what he called "competitive advantage" by adding value within their organization. Subsequently, the term was adopted for agricultural development purposes and has now become very much in vogue among those working in this field, with an increasing number of bilateral and multilateral aid organisations using it to guide their development interventions.

At the heart of the agricultural value chain concept is the idea of actors connected along a chain producing and delivering goods to consumers through a sequence of activities. However, this "vertical" chain cannot function in isolation and an important aspect of the value chain approach is that it also considers "horizontal" impacts on the chain, such as input and finance provision, extension support and the general enabling environment. The approach has been found useful, particularly by donors, in that it has resulted in a consideration of all those factors impacting on the ability of farmers to access markets profitably, leading to a broader range of chain interventions. It is used both for upgrading existing chains and for donors to identify market opportunities for small farmers.

Definitions

There is no commonly agreed definition of what is actually meant by agricultural value chains. Indeed, some agencies are using the term without having a workable definition or definitions and simply redefined ongoing activities as "value chain" work when the term came into vogue. Published definitions include the World Bank's "the term ''value chain'' describes the full range of value adding activities required to bring a product or service through the different phases of production, including procurement of raw materials and other inputs", UNIDO's "actors connected along a chain producing, transforming and bringing goods and services to end-consumers through a sequenced set of activities", and CIAT's "a strategic network among a number of business organizations".

Without a universal definition the term "value chain" is now being used to refer to a range of types of chain, including:

- An international, or regional commodity market. Examples could include "the global cotton value chain", "the southern African maize value chain" or "the Brazilian coffee value chain";
- A national or local commodity market or marketing system such as "the Ghanaian tomato value chain" or ""the Accra tomato value chain";
- A supply chain, which can cover both of the above;
- An extended supply chain or marketing channel, which embraces all activities needed to produce the product, including information/extension, planning, input supply and finance. It is probably the most common usage of the value chain term;
- A dedicated chain designed to meet the needs of one or a limited number of buyers. This usage, which is arguably most faithful to Porter's concept, stresses that a value chain is designed to capture value for all actors by carrying out activities to meet the demand of consumers or of a particular retailer, processor or food service company supplying those consumers. Emphasis is firmly placed on demand as the source of the value.

Value chain methodologies

Donors and others supporting agricultural development, such as GIZ, DFID, ILO, IIED and UNIDO, have produced a range of documents designed to assist their staff and others to evaluate value chains in order to decide on the most appropriate interventions to either update existing chains or promote new ones. However, the application of value chain analysis is being interpreted differently by different organisations, with possible repercussions for their development impact. The proliferation of guides has taken place in an environment where key conceptual and methodological elements of value chain analysis and development are still evolving. Many of these guides include not only detailed procedures that require experts to carry out the analysis but also use detailed quasi-academic methodologies. One such methodology is to compare the same value chain over time (a comparative or panel study) to assess changes in rents, governance, systemic efficiency and the institutional framework.

LINKING FARMERS TO MARKETS

A major subset of value chain development work is concerned with ways of linking producers to companies, and hence into the value chains. While there are examples of fully integrated value chains that do not involve smallholders (e.g. Unilever operates tea estates and tea processing facilities in Kenya and then blends and packs the tea in Europe before selling it as Lipton, Brooke Bond or PG Tips brands), the great bulk of agricultural value chains involve sales to companies from independent farmers. Such arrangements frequently involve contract farming in which the farmer undertakes to supply agreed quantities of a crop or livestock product, based on the quality standards and delivery requirements of the purchaser, often at a price that is established in advance. Companies often also agree to support the farmer through input supply, land preparation, extension advice and transporting produce to their premises.

Inclusive value chains

Work to promote market linkages in developing countries is often based on the concept of "inclusive value chains", which usually places emphasis on identifying possible ways in which small-scale farmers can be incorporated into existing or new value chains or can extract greater value from the chain, either by increasing efficiency or by also carrying out activities further along the chain. In the various publications on the topic the definition of "inclusion" is often imprecise as it is often unclear whether the development aim is to include all farmers or only those best able to take advantage of the opportunities.

Agricultural value chain finance

Agricultural value chain finance is concerned with the flows of funds to and within a value chain to meet the needs of chain actors for finance, to secure sales, to buy inputs or produce, or to improve efficiency. Examining the potential for value chain finance involves a holistic approach to analyze the chain, those working in it, and their inter-linkages. These linkages allow financing to flow through the chain. For example, inputs can be provided to farmers and the cost can be repaid directly when the product is delivered, without need for farmers taking a loan from a bank or similar institution. This is common under contract farming arrangements. Types of value chain finance include product financing through trader and input supplier credit or credit supplied by a marketing company or a lead firm. Other trade finance instruments include receivables financing where the bank advances funds against an assignment of future receivables from the buyer, and factoring in which a business sells its accounts receivable at a discount. Also falling under value chain finance are asset collateralization, such as on the basis of warehouse receipts, and risk mitigation, such as forward contracting, futures and insurance.

The use of ICTs in value chains

Information and Communication Technologies, or ICTs, have become an important tool in promoting agricultural value chain efficiency. There has been a rapid expansion in the use of mobile technologies, in particular. The price of ICT services is falling and the technologies are becoming more affordable to many in developing countries. Applications can support farmers directly through SMS messages. Examples include iCow, developed in Kenya, which provides information on the gestation period, on artificial insemination of the cows, and on how to look after them. Applications such as M-Pesa can support access to mobile payment services for a large percentage of those without banks, thereby facilitating transactions in the value chain. Other applications have been developed to promote provision of crop insurance through input dealers, for example.

ICTs are also being used to strengthen the capacity of extension officers and NGO field staff to reach farmers with timely and accurate information and, at the same time, help capture data from the field. The Grameen Foundation's Community Knowledge Worker (CKW) programme is an example. Farmer representatives are trained to use ICT applications on a smartphone to provide agricultural information and extension support. Most market price information is now delivered to farmers via SMS. Further along the chain, technologies offer considerable possibilities to enhance traceability, which is particularly relevant as certification grows in importance. Where necessary many exporters can now trace consignments back to individual farmers and take necessary measures to address problems. Finally, systems such as FARA's eRails are also supporting agricultural researchers through data collection and analysis and access to up-to-date research publications.

Enabling environments

As with all agricultural growth, two things appear essential for successful value chain development: creating the right environment for agriculture and investing in rural public goods. An enabling environment implies peace and public order, macro-economic stability, inflation under control, exchange rates based on market fundamentals rather than government allocation of foreign currency, predictable taxation that is reinvested in public goods and property rights. There is a positive correlation of agricultural growth with investment in irrigation, transport infrastructure and other technologies. Governments have a responsibility to provide essential goods and services, infrastructure, such as rural roads, and agricultural research and extension. Value chain development is often constrained by corruption, both at a high level and at the ubiquitous road blocks found in many countries, particularly in Africa. Many measures to improve value chains require collaboration between a wide range of different ministries, and this can be difficult to achieve.

Intensive farming

Intensive farming involves various types of agriculture with higher levels of input and output per cubic unit of agricultural land area. It is characterized by a low fallow ratio, higher use of inputs such as capital and labour, and higher crop yields per cubic unit land area.This contrasts with traditional agriculture, in which the inputs per unit land are lower. The term "intensive" involves various meanings, some of which refer to organic farming methods (such as biointensive agriculture and French intensive gardening), and others that refer to nonorganic and industrial methods. Intensive animal farming involves either large numbers of animals raised on limited land, usually concentrated animal feeding operations (CAFOs), often referred to as factory farms, or managed intensive rotational grazing(MIRG), which has both organic and non-organic types. Both increase the yields of food and fiber per acre as compared to traditional animal husbandry. In CAFO, feed is brought to the seldom-moved animals, while in MIRG the animals are repeatedly moved to fresh forage.

Most commercial agriculture is intensive in one or more ways. Forms that rely heavily on industrial methods are often called industrial agriculture, which is characterised by innovations designed to increase yield. Techniques include planting multiple crops per year, reducing the frequency of fallow years, and improving cultivars. It also involves increased use of fertilizers, plant growth regulators, and pesticides and mechanised agriculture, controlled by increased and more detailed analysis of growing conditions, including weather, soil, water, weeds, and pests. This system is supported by ongoing innovation in agricultural machinery and farming methods, genetic technology, techniques for achieving economies of scale, logistics, and data collection and analysis technology. Intensive farms are widespread in developed nations and increasingly prevalent worldwide. Most of the meat, dairy, eggs, fruits, and vegetables available in supermarkets are produced by such farms.

Smaller intensive farms usually include higher inputs of labor and more often use sustainable intensive methods. The farming practices commonly found on such farms are referred to as appropriate technology. These farms are less widespread in both developed countries and worldwide, but are growing more rapidly. Most of the food available in specialty markets such as farmers markets is produced by these small holder farms.

Agricultural development in Britain between the 16th century and the mid-19th century saw a massive increase in agricultural productivity and net output. This in turn supported unprecedented population growth, freeing up a significant percentage of the workforce, and thereby helped enable the Industrial Revolution. Historians cited enclosure, mechanization, four-field crop rotation, and selective breeding as the most important innovations.

Industrial agriculture arose along with the Industrial Revolution. By the early 19th century, agricultural techniques, implements, seed stocks, and cultivars had so improved that yield per land unit was many times that seen in the Middle Ages.

The industrialization phase involved a continuing process of mechanization. Horse-drawn machinery such as the McCormick reaperrevolutionized harvesting, while inventions such as the cotton gin reduced the cost of processing. During this same period, farmers began to use steam-powered threshers and tractors, although they were expensive and dangerous. In 1892, the first gasoline-poweredtractor was successfully developed, and in 1923, the International Harvester Farmall tractor became the first all-purpose tractor, marking an inflection point in the replacement of draft animals with machines. Mechanical harvesters (combines), planters, transplanters, and other equipment were then developed, further revolutionizing agriculture. These inventions increased yields and allowed individual farmers to manage increasingly large farms.

The identification of nitrogen, phosphorus, and potassium (NPK) as critical factors in plant growth led to the manufacture of synthetic fertilizers, further increasing crop yields. In 1909, the Haber-Bosch method to synthesize ammonium nitrate was first demonstrated. NPK fertilizers stimulated the first concerns about industrial agriculture, due to concerns that they came with serious side effects such as soil compaction, soil erosion, and declines in overall soil fertility, along with health concerns about toxic chemicals entering the food supply.

The identification of carbon as a critical factor in plant growth and soil health, particularly in the form of humus, led to so-called sustainable agriculture, as well as alternative forms of intensive agriculture that also surpassed traditional agriculture, without side effects or health issues. Farmers adopting this approach were initially referred to as humus farmers, later as organic farmers.

The discovery of vitamins and their role in nutrition, in the first two decades of the 20th century, led to vitamin supplements, which in the 1920s allowed some livestock to be raised indoors, reducing their exposure to adverse natural elements. Chemicals developed for use in World War II gave rise to synthetic pesticides.

Following World War II, synthetic fertilizer use increased rapidly, while sustainable intensive farming advanced much more slowly. Most of the resources in developed nations went to improving industrial intensive farming, and very little went to improving organic farming. Thus, particularly in the developed nations, industrial intensive farming grew to become the dominant form of agriculture.

The discovery of antibiotics and vaccines facilitated raising livestock in confined animal feeding operations by reducing diseases caused by crowding. Developments in logistics and refrigeration as well as processing technology made long-distance distribution feasible.

Between 1700 and 1980, "the total area of cultivated land worldwide increased 466%" and yields increased dramatically, particularly because of selectively-bred, high-yielding varieties, fertilizers, pesticides, irrigation, and machinery. Global agricultural production doubled between 1820 and 1920; between 1920 and 1950; between 1950 and 1965; and again between 1965 and 1975 to feed a global population that grew from one billion in 1800 to 6.5 billion in 2002. The number of people involved in farming in industrial countries dropped, from 24 percent of the American population to 1.5 percent in 2002. In 1940, each farmworker supplied 11 consumers, whereas in 2002, each worker supplied 90 consumers. The number of farms also decreased and their ownership became more concentrated. In the year

2000 in the U.S., four companies produced 81 percent of cows, 73 percent of sheep, 57 percent of pigs, and 50 percent of chickens, which was cited as an example of "vertical integration" by the president of the U.S. National Farmers Union.Between 1967 and 2002, the one million pig farms in America consolidated into 114,000, with 80 million pigs (out of 95 million) produced each year on factory farms, according to the U.S. National Pork Producers Council. According to the Worldwatch Institute, 74 percent of the world's poultry, 43 percent of beef, and 68 percent of eggs are produced this way.

Concerns over the sustainability of industrial agriculture, which has become associated with decreased soil quality, and over the environmental effects of fertilizers and pesticides, have not subsided. Alternatives such as integrated pest management (IPM) have had little impact because policies encourage the use of pesticides and IPM is knowledge-intensive. These concerns sustained the organic movement and caused a resurgence in sustainable intensive farming, as well as funding for the development of appropriate technology.

TECHNIQUES AND TECHNOLOGIES

Intensive livestock farming, also called "factory farming", is a term referring to the process of raising livestock in confinement at high stocking density. "Concentrated animal feeding operations" (CAFO), or "intensive livestock operations", can hold large numbers (some up to hundreds of thousands) of cows, hogs, turkeys, or chickens, often indoors. The essence of such farms is the concentration of livestock in a given space. The aim is to provide maximum output at the lowest possible cost and with the greatest level of food safety. The term is often used pejoratively. However, CAFOs have dramatically increased the production of food from animal husbandry worldwide, both in terms of total food produced and efficiency.

Food and water is delivered to the animals, and therapeutic use of antimicrobial agents, vitamin supplements, and growth hormones are often employed. Growth hormones are not used on chickens nor on any animal in the European Union. Undesirable behaviors often related to the stress of confinement led to a search for docile breeds (e.g., with natural dominant behaviors bred out), physical restraints to stop interaction, such as individual cages for chickens, or physical modification such as the de-beaking of chickens to reduce the harm of fighting.

The CAFO designation resulted from the 1972 U.S. Federal Clean Water Act, which was enacted to protect and restore lakes and rivers to a "fishable, swimmable" quality. The United States Environmental Protection Agency (EPA) identified certain animal feeding operations, along with many other types of industry, as "point source" groundwater polluters. These operations were subjected to regulation.

In 17 states in the U.S., isolated cases of groundwater contamination were linked to CAFOs. For example, the ten million hogs in North Carolina generate 19 million tons of waste per year. The U.S. federal government acknowledges the waste disposal issue and requires that animal waste be stored in lagoons. These lagoons can be as large as 7.5 acres (30,000 m). Lagoons not protected with an impermeable liner can leak into groundwater under some conditions, as can runoff from manure used as fertilizer. A lagoon that burst in 1995 released 25 million gallons of nitrous sludge in North Carolina's New River. The spill allegedly killed eight to ten million fish.

The large concentration of animals, animal waste, and dead animals in a small space poses ethical issues to some consumers. Animal rights and animal welfare activists have charged that intensive animal rearing is cruel to animals.

Other concerns include persistent noxious odor, the effects on human health, and the role of antibiotic use in the rise of resistant infectious bacteria.

According to the U.S. Centers for Disease Control and Prevention (CDC), farms on which animals are intensively reared can cause adverse health reactions in farm workers. Workers may develop acute and/or chronic lung disease, musculoskeletal injuries, and may catch (zoonotic) infections from the animals.

Managed intensive rotational grazing

Managed Intensive Rotational Grazing (MIRG), also known as cell grazing, mob grazing, and holistic management planned grazing, is a variety of foraging in which herds or flocks are regularly and systematically moved to fresh, rested grazing areas to maximize the quality and quantity of forage growth. MIRG can be used with cattle, sheep, goats, pigs, chickens, turkeys, ducks, and other animals. The herds graze one portion of pasture, or a paddock, while allowing the others to recover. Resting grazed lands allows the vegetation to renew energy reserves, rebuild shoot systems, and deepen root systems, resulting in long-term maximum biomass production. MIRG is especially effective because grazers thrive on the more tender younger plant stems. MIRG also leaves parasites behind to die off, minimizing or eliminating the need for de-wormers. Pasture systems alone can allow grazers to meet their energy requirements, and with the increased productivity of MIRG systems, the animals obtain the majority of their nutritional needs, in some cases all, without the supplemental feed sources that are required in continuous grazing systems.

Pasture intensification

Pasture intensification is the improvement of pasture soils and grasses to increase the food production potential of livestock systems. It is commonly used to reverse pasture degradation, a process characterized by loss of forage and decreased animal carrying capacity which results from overgrazing, poor nutrient management, and lack of soil conservation. This degradation leads to poor pasture soils with decreased fertility and water availability and increased rates of erosion, compaction, and acidification. Degraded pastures have significantly lower productivity and higher carbon footprints compared to intensified pastures.

Management practices which improve soil health and consequently grass productivity include irrigation, soil scarification, and the application of lime, fertilizers, and pesticides. Depending on the productivity goals of the target agricultural system, more involved restoration projects can be undertaken to replace invasive and under-productive grasses with grass species that are better suited to the soil and climate conditions of the region. These intensified grass systems allow higher stocking rates with faster animal weight gain and reduced time to slaughter, resulting in more productive, carbon-efficient livestock systems.

Another technique to optimize yield while maintaining the carbon balance is the use of integrated crop-livestock (ICL) and crop-livestock-forestry (ICLF) systems, which combine several ecosystems into one optimized agricultural framework. These synergies between these systems provide benefits to pastures through optimal plant usage, improved feedand fattening rates, increased soil fertility and quality, intensified nutrient cycling, integrated pest control, and improved biodiversity. The introduction of certain legume crops to pastures increases carbon accumulation and nitrogen fixation in soils, while their digestibility helps animal fattening and reduces methane emissions from enteric fermentation.ICLF systems yield beef cattle productivity up to ten times that of degraded pastures, additional crop production from maize, sorghum, and soybean harvests, and greatly reduced greenhouse gas balances due to forest carbon sequestration.

Crops

The Green Revolution transformed farming in many developing countries. It spread technologies that had already existed, but had not been widely used outside of industrialized nations. These technologies included "miracle seeds", pesticides, irrigation, and synthetic nitrogen fertilizer.

Seeds

In the 1970s, scientists created strains of maize, wheat, and rice that are generally referred to as high-yielding varieties (HYV). HYVs have an increased nitrogen-absorbing potential compared to other varieties. Since cereals that absorbed extra nitrogen would typically lodge (fall over) before harvest, semi-dwarfing genes were bred into their genomes. Norin 10 wheat, a variety developed by Orville Vogel from Japanese dwarf wheat varieties, was instrumental in developing wheat cultivars. IR8, the first widely implemented HYV rice to be developed by the International Rice Research Institute, was created through a cross between an Indonesian variety named "Peta" and a Chinese variety named "Dee Geo Woo Gen."

With the availability of molecular genetics in Arabidopsis and rice the mutant genes responsible (reduced height (rht), gibberellin insensitive (gai1) and slender rice (slr1)) have been cloned and identified as cellular signalling components of gibberellic acid, a phytohormone involved in regulating stem growth via its effect on cell division. Photosynthetic investment in the stem is reduced dramatically as the shorter plants are inherently more mechanically stable. Nutrients become redirected to grain production, amplifying in particular the yield effect of chemical fertilizers.

HYVs significantly outperform traditional varieties in the presence of adequate irrigation, pesticides, and fertilizers. In the absence of these inputs, traditional varieties may outperform HYVs. They were developed as F1 hybrids, meaning seeds need to be purchased every season to obtain maximum benefit, thus increasing costs.

Crop rotation

Crop rotation or crop sequencing is the practice of growing a series of dissimilar types of crops in the same space in sequential seasons for benefits such as avoiding pathogen and pest buildup that occurs when one species is continuously cropped. Crop rotation also seeks to balance the nutrient demands of various crops to avoid soil nutrient depletion. A traditional component of crop rotation is the replenishment of nitrogen through the use of legumes and green manure in sequence with cereals and other crops. Crop rotation can also improve soil structure and fertility by alternating deep-rooted and shallow-rooted plants. A related technique is to plant multi-species cover crops between commercial crops. This combines the advantages of intensive farming with continuous cover and polyculture.

Irrigation

Crop irrigation accounts for 70% of the world's fresh water use. Flood irrigation, the oldest and most common type, is typically unevenly distributed, as parts of a field may receive excess water in order to deliver sufficient quantities to other parts. Overhead irrigation, using center-pivot or lateral-moving sprinklers, gives a much more equal and controlled distribution pattern. Drip irrigation is the most expensive and least-used type, but delivers water to plant roots with minimal losses.

Water catchment management measures include recharge pits, which capture rainwater and runoff and use it to recharge groundwater supplies. This helps in the replenishment of groundwater wells and eventually reduces soil erosion. Dammed rivers creating reservoirs store water for irrigation and other uses over large areas. Smaller areas sometimes use irrigation ponds or groundwater.

Weed control

In agriculture, systematic weed management is usually required, often performed by machines such as cultivators or liquid herbicide sprayers. Herbicides kill specific targets while leaving the crop relatively unharmed. Some of these act by interfering with the growth of the weed and are often based on plant hormones. Weed control through herbicide is made more difficult when the weeds become resistant to the herbicide. Solutions include:

- Cover crops (especially those with allelopathic properties) that out-compete weeds or inhibit their regeneration
- Multiple herbicides, in combination or in rotation
- Strains genetically engineered for herbicide tolerance
- Locally adapted strains that tolerate or out-compete weeds
- Tilling
- Ground cover such as mulch or plastic

- Manual removal
- Mowing
- Grazing
- Burning

TERRACING

In agriculture, a terrace is a leveled section of a hilly cultivated area, designed as a method of soil conservation to slow or prevent the rapid surface runoff of irrigation water. Often such land is formed into multiple terraces, giving a stepped appearance. The human landscapes of rice cultivation in terraces that follow the natural contours of the escarpments, like contour ploughing, are a classic feature of the island of Bali and the Banaue Rice Terraces in Banaue, Ifugao, Philippines. In Peru, the Inca made use of otherwise unusable slopes by building drystone walls to create terraces.

Rice paddies

A paddy field is a flooded parcel of arable land used for growing rice and other semiaquatic crops. Paddy fields are a typical feature of rice-growing countries of east and southeast Asia, including Malaysia, China, Sri Lanka, Myanmar, Thailand, Korea, Japan, Vietnam, Taiwan, Indonesia, India, and the Philippines. They are also found in other rice-growing regions such as Piedmont (Italy), the Camargue (France), and the Artibonite Valley (Haiti). They can occur naturally along rivers or marshes, or can be constructed, even on hillsides. They require large water quantities for irrigation, much of it from flooding. It gives an environment favourable to the strain of rice being grown, and is hostile to many species of weeds. As the only draft animal species which is comfortable in wetlands, the water buffalo is in widespread use in Asian rice paddies.

Paddy-based rice-farming has been practiced in Korea since ancient times. A pit-house at the Daecheon-ni archaeological site yielded carbonized rice grains and radiocarbon dates indicating that rice cultivation may have begun as early as the Middle Jeulmun Pottery Period (c. 3500-2000 BC) in the Korean Peninsula. The earliest rice cultivation there may have used dry-fields instead of paddies.

The earliest Mumun features were usually in naturally swampy, low-lying narrow gulleys and fed by local streams. Some Mumun paddies in flat areas were made of a series of squares and rectangles separated by bunds approximately 10 cm in height, while terraced paddies were long and irregular in shape, following the natural contours of the land at various levels.

Like today, Mumun period rice farmers used terracing, bunds, canals, and small reservoirs. Some paddy-farming techniques of the Middle Mumun period (c. 850-550 BC) can be interpreted from the well-preserved wooden tools excavated from archaeological rice paddies at the Majeon-ni site. Iron tools for paddy-farming were not introduced until sometime after 200 BC. The spatial scale of individual paddies, and thus entire paddy-fields, increased with the regular use of iron tools in the Three Kingdoms of Korea Period (c.AD 300/400-668). A recent development in the intensive production of rice is System of Rice Intensification (SRI). Developed in 1983 by the French Jesuit Father Henri de Laulanié in Madagascar,by 2013 the number of smallholder farmers using SRI had grown to between 4 and 5 million.

Aquaculture

Aquaculture is the cultivation of the natural products of water (fish, shellfish, algae, seaweed, and other aquatic organisms). Intensive aquaculture takes place on land using tanks, ponds, or other controlled systems, or in the ocean, using cages.

Sustainable intensive farming

Sustainable intensive farming practices have been developed to slow the deterioration of agricultural land and even regenerate soil health and ecosystem services, while still offering high yields. Most of these developments fall in the category of organic farming, or the integration of organic and conventional agriculture. "Organic systems and the practices that make them effective are being picked up more and more by conventional agriculture and will become the foundation for future farming systems. They won't be called organic, because they'll still use some chemicals and still use some fertilizers, but they'll function much more like today's organic systems than today's conventional systems."

Dr. Charles Benbrook Executive director US House Agriculture Subcommittee Director Agricultural Board - National Academy Sciences (FMR) The System of Crop Intensification (SCI) was born out of research primarily at Cornell University and smallholder farms in India on SRI. It uses the SRI concepts and methods for rice and applies them to crops like wheat, sugarcane, finger millet, and others. It can be 100% organic, or integrated with reduced conventional inputs.

Holistic management is a systems thinking approach that was originally developed for reversing desertification. Holistic planned grazing is similar to rotational grazing but differs in that it more explicitly provides a framework for adapting to four basic ecosystem processes: the water cycle, the mineral cycle (including the carbon cycle), energy flow, and community dynamics (the relationship between organisms in an ecosystem) as equal in importance to livestock production and social welfare. By intensively managing the behavior and movement of livestock, holistic planned grazing simultaneously increases stocking rates and restores grazing land.

Pasture cropping involves planting grain crops directly into grassland without first applying herbicides. The perennial grasses form a living mulch understory to the grain crop, eliminating the need to plant cover crops after harvest. The pasture is intensively grazed both before and after grain production using holistic planned grazing. This intensive system yields equivalent farmer profits (partly from increased livestock forage) while building new topsoil and sequestering up to 33 tons of CO_2/ha/year.

The Twelve Aprils grazing program for dairy production, developed in partnership with the USDA-SARE, is similar to pasture cropping, but the crops planted into the perennial pasture are forage crops for dairy herds. This system improves milk production and is more sustainable than confinement dairy production.

Integrated multi-trophic aquaculture (IMTA) is an example of a holistic approach. IMTA is a practice in which the by-products (wastes) from one species are recycled to become inputs (fertilizers, food) for another. Fed aquaculture (e.g. fish, shrimp) is combined with inorganic extractive (e.g. seaweed) and organic extractive aquaculture (e.g. shellfish) to create balanced systems for environmental sustainability (biomitigation), economic stability (product diversification and risk reduction), and social acceptability (better management practices).

Biointensive agriculture focuses on maximizing efficiency such as per unit area, energy input, and water input.

Agroforestry combines agriculture and orchard/forestry technologies to create more integrated, diverse, productive, profitable, healthy, and sustainable land-use systems.

Intercropping can increase yields or reduce inputs and thus represents (potentially sustainable) agricultural intensification. However, while total yield per acre is often increased dramatically, yields of any single crop often diminish. There are also challenges to farmers relying on farming equipment optimized for monoculture, often resulting in increased labor inputs.

Vertical farming is intensive crop production on a large scale in urban centers, in multi-story, artificially-lit structures, using far less inputs and producing fewer environmental impacts.

An integrated farming system is a progressive, biologically-integrated sustainable agriculture system such as IMTA or Zero waste agriculture, whose implementation requires exacting knowledge of the interactions of multiple species and whose benefits include sustainability and increased profitability. Elements of this integration can include:

- Intentionally introducing flowering plants into agricultural ecosystems to increase pollen-and nectar-resources required by natural enemies of insect pests
- Using crop rotation and cover crops to suppress nematodes in potatoes

Challenges

[W]hen hunter-gatherers with growing populations depleted the stocks of game and wild foods across the Near East, they were forced to introduce agriculture. But agriculture brought much longer hours of work and a less rich diet than hunter-gatherers enjoyed. Further population growth among shifting slash-and-burn farmers led to shorter fallow periods, falling yields and soil erosion. Plowing and fertilizers were introduced to deal with these problems - but once again involved longer hours of work and degradation of soil resources (Boserup, The Conditions of

Agricultural Growth, Allen and Unwin, 1965, expanded and updated in Population and Technology, Blackwell, 1980.).

While the point of industrial agriculture is to profitably supply the world at the lowest cost, industrial methods have significant side effects. Further, industrial agriculture is not an indivisible whole, but instead is composed of multiple elements, each of which can be modified in response to market conditions, government regulation, and further innovation, and has its own side-effects. Various interest groups reach different conclusions on the subject.

Population growth

- 30,000 years ago hunter-gatherer behavior fed 6 million people
- 3,000 years ago primitive agriculture fed 60 million people
- 300 years ago intensive agriculture fed 600 million people
- Today industrial agriculture attempts to feed 8 billion people

Between 1930 and 2000, U.S. agricultural productivity (output divided by all inputs) rose by an average of about 2 percent annually, causing food prices to decrease. "The percentage of U.S. disposable income spent on food prepared at home decreased, from 22 percent as late as 1950 to 7 percent by the end of the century."

Other impacts

Industrial agriculture uses huge amounts of water, energy, and industrial chemicals, increasing pollution in the arable land, usable water, and atmosphere. Herbicides, insecticides, and fertilizers are accumulating in ground and surface waters. "Many of the negative effects of industrial agriculture are remote from fields and farms. Nitrogen compounds from the Midwest, for example, travel down the Mississippi to degrade coastal fisheries in the Gulf of Mexico. But other adverse effects are showing up within agricultural production systems-for example, the rapidly developing resistance among pests is rendering our arsenal of herbicides and insecticides increasingly ineffective."Agrochemicals and monoculture have been implicated in Colony Collapse Disorder, in which the individual members of bee colonies disappear. Agricultural production is highly dependent on bees to pollinate many varieties of fruits and vegetables.

Social

A study done for the U.S. Office of Technology Assessment conducted by the UC Davis Macrosocial Accounting Project concluded that industrial agriculture is associated with substantial deterioration of human living conditions in nearby rural communities.

Land banking

Land banking is the practice of aggregating parcels of land for future sale or development. While in many countries land banking may refer to various private real-estate investment schemes, in the United States it refers to the establishment of quasi-governmental county or municipal authorities purposed with managing an inventory of surplus land.

MUNICIPAL LAND BANKS IN THE UNITED STATES

Land banks are quasi-governmental entities created by counties or municipalities to effectively manage and repurpose an inventory of underused, abandoned, or foreclosed property. They are often chartered to have powers that enable them to accomplish these goals in ways that existing government agencies can not. While the land bank "model" has gained broad support and has been implemented in a number of cities, they are implemented differently so as to best address both municipal needs and the state and local legal context in which they were created.

History

The period of deindustrialization in the United States coupled with increased suburbanization in the middle of the 20th century left many American cities with large amounts of vacant & blighted industrial, residential, and commercial property. Beginning in the early 1970s municipalities began to seek solutions to manage decline or spur revitalization in once prosperous city neighborhoods. The first land bank was created in St.

Louis in 1971. While additional municipalities continued to adopt them at a trickle it wasn't until the mid 2000s that land banks became viewed as a tested, reliable, and accepted model and experienced widespread implementation - particularly after the success of the Genesee County Land Bank. In 2009 the Department of Housing and Urban Development issued a report embracing land banks as a best practices model for municipalities dealing with the effects of the real estate market collapse and the ensuing foreclosure crisis.

Investment land banking by country

Land banking in the UK was previously the preserve of the landed gentry or real estate developers. Many reputable listed commercial building companies engage successfully in land banking for future building projects. Companies also purchase land sites and easily divide them into smaller plots, then offer these plots for sale to individual investors. This relatively new practice in the UK does not fall under the control of the Financial Services Authority. Many people are wary of this form of investment as many plot based land banking companies have failed or been closed down. There are currently no audited successes recorded for UK plot based land banking despite the UK having gone through a major property boom in the period 2002 - Jul 2007.

A landbanking scheme that is a Collective investment scheme is a "regulated activity" for the purposes of the Financial Services and Markets Act 2000 and, according to section 19(1), may only be operated in the UK by a person who is either authorised or exempt. Section 26 provides that an agreement made by a person in contravention of this is unenforceable and any sums paid to him may be recovered together with compensation for any loss suffered. After recent FSA enforcement of this regulation many companies selling UK land plots have moved outside of the European Union and only offer land plots to non UK residents who are not protected by FSA regulations.

Companies offering land banking plots in the UK

Since the changes in the land registration act, a number of companies offering UK land plots as an investment have been formed. Typically this land is greenbelt, nature conservation, flood plain, agricultural or protected land unsuitable for development. There are no recorded successful planning permission applications for plots sold under such collective investment schemes.

There have been considerable losses recorded by investors in UK land plot investment schemes. A large number of British companies offering UK land plots have failed or been shut down by the FSA or other authorities. Some companies have now moved offshore after FSA investigation. Some companies now offer UK land plots from locations such as Dubai or Singapore where the local authorities do not regulate such activities, or are not aware of the high risk nature of the investment. In June 2010 the Monetary Authority of Singapore (MAS) issued a warning on land banking plots schemes warning they may be a scam with specific focus on companies offering land from the UK and Canada.

Sales methods

A company representative may contact an individual by telephone, in temporary shopping center booths, or at property shows and offer a strategic land investment in the UK. Very often UK government or industry statistics, the proximity of the land to built up areas, or the recent history of UK house prices are quoted as a demonstration of why the land plot is a great investment. Verbal communication will often indicate that the land is fast tracked for building approval and has strong potential as building land. When pricing the land reference is typically made to approved building land prices at the market peak. Very often the land banking company will present detailed plans showing a housing development on the site. These plans are often referred to as "pre-approved", "concept" or "predevelopment". The sales person will focus on the potential future value of the land against the current selling price.

No reference is ever made to the value of green belt or agricultural land, or the issues involved with the long term maintenance, and collectively selling tiny plots of land. The sales price is typically uplifted 10-100 times over the current value of the land. Plans shown have no validity in UK planning law and cannot be considered as any indication of progress in the planning process. No written contractual promise is ever given for planning permission despite typically extreme optimism by the sales person. The sales person will typically never mention that the land is protected, or greenbelt land and cannot be developed under current planning regulations. There is typically no possibility of planning permission in any reasonable timeframe.

The investor may end up paying a considerable amount of money for a small area of low value land which has a very high risk of standing undeveloped. Once the general public becomes aware of the lack of viability of the proposed plot investment scheme, the re-sale value of the individual plots collapses. This is typically followed by the land plot company liquidating completely, or relocating to another legal jurisdiction.

For customers that show a willingness to purchase such schemes there may also be attempts to sell additional plot based land banking products at alternate locations, or other high yield investment programmes. Customers may also be added to suckers lists which are then sold to other companies offering similar schemes. When the land banking plot company fails plot investors may also be offered investment recovery or planning services for a fee. Such services typically are fraudulent or fail and lead to a further loss of money for the investor.

Controversies

A You and Yours documentary, first aired on BBC Radio 4 in December 2006, criticized the services offered by many land banking companies in the United Kingdom, suggesting that they were scamming their customers. A land banking scam is based on the very low chance of any of the plots receiving planning permission and the very high profit margins taken on the land plots, with the seller using misleading marketing tactics to convince the buyers that they are making a sound investment.

A key strategy used for selling United Kingdom land plots is to imply that because a customer owns the land plot, they cannot lose their money. The land banking company typically suggests dramatic annual increases in the value of the land plots, and a very optimistic time frame for successful planning applications. These are never contractually committed. Typically the land banking company sells a land plot at a premium of 15 to 100 times the current market value of undeveloped land. A purchaser might pay £15,000 for a land plot that only has a current market value of £500. On this basis, most of the investment is not in land, and the small percentage annual increases in the value of the land plot are meaningless. The actual investment is in a proposed service to deliver valuable approved building land in the future. If that service is never delivered or is not successful, the remaining land asset is normally worthless. Should the selling company fail or disappear, the plot owner cannot economically sell the plot, as the administrative effort and cost of sale typically exceeds the value of the land plot.

Many land banking companies target victims outside of the United Kingdom, particularly Canada, Singapore, Thailand, Brunei and Malaysia. Residents of these countries may be naive of the UK property market and local planning regulations such as green belt zoning.

In 2008, the land banking firm UKLI was placed into administration due to insolvency, despite having taken £69 million from 4,500 people for land plots. Land International was closed down in 2008 after losing investors £10 million, and the same Land International plots were later offered for sale in Asia. In 2010, Land International (Far East) failed, causing investors to lose S$6 million (£2.5M). MP David Heath requested a debate in the House of Commons following the offering of 209 plots in the village of Dean, saying that "while land banking may not be illegal it is undoubtedly a scam".

The UK Land Registry issued a press release on January 15, 2009 advising consumers that the Land Registry has published a guide warning against land banking investment schemes. Land Registry Head of Corporate Legal Services Mike Westcott Rudd said that the public were being "misled about the prospects of obtaining planning permission", with well-known banks and developers being falsely cited as partners in the project, and that in some cases forged Land Registry paperwork was being presented to suggest that planning approval existed where it did not.

United States

In 2011 New York State passed a land bank statute authorizing the establishment of nonprofits in each county to take title to vacant abandoned homes so they can be rehabilitated, sold or demolished in an orderly fashion. Many counties upstate including Erie County, Onondaga County, Schenectady County and Albany County have abandoned homes as people moved to the suburbs. Some properties have been abandoned due to back taxes and the city has taken title. The recent robo-signing settlement gave Attorney General Eric Schneiderman the wherewithal to fund land banks in Schenectady and Albany. The state of Michigan also has a land bank program. Ohio passed land bank legislation in 2009.

AGRICULTURAL MARKETING IN INDIA

India is an agricultural country and one third population depends on the agricultural sector directly or indirectly. Agriculture remains as the main stray of the Indian economy since times immemorial. Indian agriculture contribution to the national gross domestic product (GDP) is about 25 per cent. With food being the crowning need of mankind, much emphasis has been on commercialising agricultural production. For this reason, adequate production and even distribution of food has of late become a high priority global concern.

Agricultural marketing is mainly the buying and selling of agricultural products. In earlier days when the village economy was more or less self-sufficient the marketing of agricultural products presented no difficulty as the farmer sold his produce to the consumer on a cash or barter basis. Agricultural Marketing, rather than production, is going to be the key driver of the agriculture sector today, thanks to the new market realities posed by the increasing accent on globalization, liberalization and privatization of the economy. Market-driven production is an idea whose time has come. With the gradual shifting of agricultural system from subsistence to commercial one, there is increasing focus on Agripreneurship and Agri-marketing. It is the need of the time to tune up the Agricultural Marketing System of the country to enable the farmers to face the new challenges and reap the opportunities as well. This summons us to revisit our traditional statistic policies and laws and bring about the requisite reforms in the sector.

The imperatives of the integration of the Agricultural Marketing System of the country are increasingly posing new challenges in respect of each of the components of the system such as- cleaning, grading, quality certification, packaging, storage, transportation, financing, wholesaling and retailing etc of agriculture produce. The subtlety of the management of Agricultural Marketing issues today calls for high degree of professionalism to measure up to the expectations of different stakeholders.

Internal reforms in traditional agricultural marketing system of the country have become indispensable to enable our farmers to tap into the external market of the post-WTO regime. Now that many states have introduced some reforms in agricultural marketing, the need of the hour is to consolidate the gain of reforms through appropriate policies and plans. The existing trade barriers have to be removed for evolving a common market in India. There is a need to give a push to processing and value addition. Cost effectiveness through integrated supply chain management has become sine qua non in the present world trade environment. This demands doing away with the dysfunctional aspects of the present regulated marketing system by putting in place a responsive market information system, need-based price discovery mechanism, measures to solve the problem of economies of scale at the grass root level by promoting contract farming, direct marketing and private investment in agricultural marketing infrastructure.

Congenial investor-friendly economic environment has got to be evolved to attract the private investors to invest in agricultural marketing sector for bridging the existing gap in the agricultural marketing infrastructure of the country. Farmers should get freedom of choice to use a physical market giving them services for the fees paid by them. This could be possible through dismantling of the monopoly of the Government-owned markets and permitting setting up markets in the private sector, thereby bringing healthy competition in the sector. This also calls for the traditional Government mandies to be spruced up in terms of their management, infrastructure, service delivery, customer-friendliness through fine-tuning of their internal processes. The other areas of reforms such as promotion of grading and standardization, market-led extension, rationalization of market fees, public- private partnership in owning and management of markets etc will go a long way towards pushing the system to the next level of excellence. The focus on reforms calls for a paradigm shift in our approach.

Today's agricultural marketing has to undergo a series of exchanges or transfers from one person to another before it reaches the consumer. There are three marketing functions involved in this, i.e., assembling, preparation for consumption and distribution. Selling on any agricultural produce depends on some couple of factors like the demand of the product at that time, availability of storage etc. The products may be sold directly in the market or it may be stored locally for the time being. Moreover, it may be sold as it is gathered from the field or it may be cleaned, graded and processed by the farmer or the merchant of the village. Sometime processing is done because consumers want it, or sometimes to conserve the quality of that product. The task of distribution system is to match the supply with the existing demand by whole selling and retailing in various points of different markets like primary, secondary or terminal markets.

Most of the agricultural products in India are sold by farmers in the private sector to moneylenders (to whom the farmer may be indebted) or to village traders. Products are sold in various ways. For example, it might be sold at a weekly village market in the farmer's village or in a neighboring village. If these outlets are not available, then produce might be sold at irregularly held markets in a nearby village or town, or in the mandi.

In India, there are several central government organisations, who are involved in agricultural marketing like, Commission of Agricultural Costs and Prices, Food Corporation of India, Cotton Corporation of India, Jute Corporation of India, etc. There are also specialised marketing bodies for rubber, tea, coffee, tobacco, spices and vegetables.

Under the Agricultural Produce (grading and marketing) Act of 1937, more than forty primary commodities are compulsorily graded for export and voluntarily graded for internal consumption. Although the regulation of commodity markets is a function of state government, the directorate of marketing and inspection provides marketing and inspection services and financial aid down to the village level to help set up commodity grading centers in selected markets.

As we have a tradition of agricultural production, marketing and allied commercial activities, now it is the time for us to brainstorm and come out with new ideas of value added services. These value added services will give the existing agricultural engine a new dimension. The next logical step could be food-processing which not only could be another revenue generating area but also can provide lots of full-time employment to our youths. With the changing agricultural scenario and global competition, there is a need of exploiting the available resources at maximum level.

There was a survey undertaken by the directorate of marketing and inspection in the ministry of agriculture in 1970-71 and 1971-72, of five hundred regulated markets was, with a view to assessing the adequacy and efficiency of the existing regulated markets and highlighting their drawbacks and deficiencies and suggesting measures to develop them. One of the most important drawbacks has been the inadequate financial resources of some of the market committees. During the fourth plan, a central sector scheme was drawn up by the ministry of agriculture to provide a grant at 20 per cent of the cost of development of market, subject to a maximum of Rs. 2 lakhs. The balance will have to be provided by the commercial banks.

Another important development in the field of regulated markets is the keen interest taken by the International Development Agency (IDA) in the development of the infrastructure in regulated markets. The IDA is financing the development of infrastructure in 50 markets of Bihar.

There are also some good news on the front of irrigation, rural infrastructure, restoring water bodies and water harvesting. Another action initiated to improve the governance of the Small Farmers Agri-business Consortium (SFAC) including the appointment of a banker as the chief executive; necessary additional capital to be provided to SFAC.

Agricultural Marketing is an area for the "second generation" of green revolution problems. Indian Marketing is undergoing a significant metamorphosis because of economic liberalization and globalisation. Infrastructure development is the critical factor for determining the success of market-oriented strategy and macro-economic policies in developing countries. Both national and global players are trying to capture the urban markets of India, which has already reached a saturation level, and it is extremely difficult to tap the urban markets with high profit margin. In export market, a firm has to face challenges from Multi-National Corporates (MNCs), other foreign companies and domestic firms of the concerned export markets. While urban market or export market is difficult to plough, rural market is relatively easy and feasible to cultivate, in India.

Market information is an important aspect of Agricultural Marketing. The importance of sound agricultural marketing policies for ensuring fair returns to the farmers can hardly be over-emphasized. It, therefore, becomes necessary on the part of regulatory agencies to ensure remunerative prices to the farmers for the sale of their produce, to boost up their efforts for increasing and sustaining the agricultural production. A number of measures have been taken by the Government to protect and safeguard the interests of farmers, like regulation of markets, grading of agricultural produce, cooperative marketing etc. Still the benefits are not percolating down to the farmers, as they are unable to plan their strategies for sale of their produce at remunerative prices, in the absence of correct and timely market information and advice about arrivals, prices, market trend, etc.

Almost all the States and Union Territories are providing market information in one form or the other for the benefits of market users like producers, traders and consumers. However, the information is collected and disseminated by use of conventional methods which cause inordinate delay in communicating the information to different target groups, and thus adversely affecting their economic interest. Therefore, there is an urgent need to bring improvement in the present market information system by linking (i) all Agricultural Produce Wholesale markets in the States and Union Territories, and (ii) the State Agricultural Marketing Boards and State Directorates of Agricultural Marketing, with the Directorate of Marketing &

Inspection of the Ministry of Agriculture, for effective and efficient information exchange. In view of this importance, the Department of Agriculture & Cooperation, in the Union Ministry of Agriculture has sanctioned a Central Sector Scheme "NICNET Based Agricultural Marketing Information Network (AGMARK-NET)", Project Cost estimate of about Rs.10.00 Crores, to link 810 nodes (735 agricultural markets and 75 State Marketing Boards/DMI offices) to begin with, through its attached office, Directorate of Marketing and Inspection (DMI), during the year 2000-02. During the X five year plan, further expansion to 2000 additional markets, Regional Portals and GIS based National atlas of Markets has also been approved . The Scheme is being implemented during the XI Plan period by enhancing its scope to cover 500 additional markets including remaining principal regulated markets, if any, farmer/consumer markets, private markets, panchayat markets, urban markets and special commodity markets.

Introduction to Agricultural Marketing

Agricultural marketing covers the services involved in moving an agricultural product from the farm to the consumer. Numerous interconnected activities are involved in doing this, such as planning production, growing and harvesting, grading, packing, transport, storage, agro- and food processing, distribution, advertising and sale. Some definitions would even include "the acts of buying supplies, rentng equipment, (and) paying labor", arguing that marketing is everything a business does. Such activities cannot take place without the exchange of information and are often heavily dependent on the availability of suitable finance.

Marketing systems are dynamic; they are competitive and involve continuous change and improvement. Businesses that have lower costs, are more efficient, and can deliver quality products, are those that prosper. Those that have high costs, fail to adapt to changes in market demand and provide poorer quality are often forced out of business. Marketing has to be customer-oriented and has to provide thefarmer, transporter, trader, processor, etc. with a profit. This requires those involved in marketing chains to understand buyer requirements, both in terms of product and business conditions.

In Western countries considerable agricultural marketing support to farmers is often provided. In the USA, for example, the USDAoperates the Agricultural Marketing Service. Support to developing countries with agricultural marketing development is carried out by various donor organizations and there is a trend for countries to develop their own Agricultural Marketing or Agribusiness units, often attached to ministries of agriculture. Activities include market information development, marketing extension, training in marketing and infrastructure development. Since the 1990s trends have seen the growing importance of supermarkets and a growing interest in contract farming, both of which impact significantly on the way in which marketing takes place.

Agricultural marketing support

In the United States the Agricultural Marketing Service (AMS) is a division of USDA and has programs for cotton, dairy, fruit and vegetable, livestock and seed, poultry, and tobacco. These programs provide testing, standardization, grading and market news services and oversee marketing agreements and orders, administer research and promotion programs, and purchase commodities for federal food programs. The AMS also enforces certain federal laws. USDA also provides support to the Agricultural Marketing Resource Center at Iowa State University and toPenn State University.

In the United Kingdom support for marketing of some commodities was provided before and after the Second World War by boards such as the Milk Marketing Board and the Egg Marketing Board, but these were closed down in the 1970s. As a colonial power Britain established marketing boards in many countries, particularly in Africa. Some continue to exist although many were closed down at the time of the introduction of structural adjustment measures in the 1990s.

In recent years several developing countries have established government-sponsored marketing or agribusiness units. South Africa, for example, started the National Agricultural Marketing Council (NAMC) as a response to the deregulation of the agriculture industry and closure of marketing boards in the country. India has the long-established National Institute of Agricultural Marketing (NIAM). These are primarily research and policy organizations, but other agencies provide facilitating services for marketing channels, such as the provision of infrastructure, market information and documentation support. Examples include the National Agricultural Marketing Development Corporation (NAMDEVCO) in Trinidad and Tobago and the New Guyana Marketing Corporation.

Several organizations provide support to developing countries to develop their agricultural marketing systems, including FAO's agricultural marketing unit and various donor organizations. There has also recently been considerable interest by NGOs to carry out activities to link farmers to markets.

Agricultural marketing development

Well-functioning marketing systems necessitates a strong private sector backed up by appropriate policy and legislative frameworks and effective government support services. Such services can include provision of market infrastructure, supply of market information (as done by USDA, for example), and agricultural extension services able to advise farmers on marketing. Training in marketing at all levels is also needed. One of many problems faced in agricultural marketing in developing countries is the latent hostility to the private sector and the lack of understanding of the role of the intermediary. For this reason "middleman" has become very much a pejorative word.

Agricultural advisory services and the market

Promoting market orientation in agricultural advisory services aims to provide for the sustainable enhancement of the capabilities of the rural poor to enable them to benefit from agricultural markets and help them to adapt to factors which impact upon these. As a study by the Overseas Development Institute demonstrates, a value chain approach to advisory services indicates that the range of clients serviced should go beyond farmers to include input providers, producers, producer organisations and processors and traders.

Market infrastructure

Efficient marketing infrastructure such as wholesale, retail and assembly markets and storage facilities is essential for cost-effective marketing, to minimise post-harvest losses and to reduce health risks. Markets play an important role in rural development, income generation, food security, developing rural-market linkages and gender issues. Planners need to be aware of how to design markets that meet a community's social and economic needs and how to choose a suitable site for a new market. In many cases sites are chosen that are inappropriate and result in under-use or even no use of the infrastructure constructed. It is also not sufficient just to build a market: attention needs to be paid to how that market will be managed, operated and maintained. In most cases, where market improvements were only aimed at infrastructure upgrading and did not guarantee maintenance and management, most failed within a few years.

Rural assembly markets are located in production areas and primarily serve as places where farmers can meet with traders to sell their products. These may be occasional (perhaps weekly) markets, such as haat bazaars in India and Nepal, or permanent. Terminal wholesale markets are located in major metropolitan areas, where produce is finally channelled to consumers through trade between wholesalers and retailers, caterers, etc. The characteristics of wholesale markets have changed considerably as retailing changes in response to urban growth, the increasing role of supermarkets and increased consumer spending capacity. These changes require responses in the way in which traditional wholesale markets are organized and managed.

Retail marketing systems in western countries have broadly evolved from traditional street markets through to the modern hypermarket or out-of-town shopping centre. In developing countries, there remains considerable scope to improve agricultural marketing by constructing new retail markets, despite the growth of supermarkets, although municipalities often view markets as sources of revenue rather than infrastructure requiring development. Effective regulation of markets is essential. Inside the market, both hygiene rules and revenue collection activities have to be enforced. Of equal importance, however, is the maintenance of order outside the market. Licensed traders in a market will not be willing to cooperate in raising standards if they face competition from unlicensed operators outside who do not pay any of the costs involved in providing a proper service.

Market information

Efficient market information can be shown to have positive benefits for farmers and traders. Up-to-date information on prices and other market factors enables farmers to negotiate with traders and also facilitates spatial distribution of products from rural areas to towns and between markets. Most governments in developing countries have tried to provide market information services to farmers, but these have tended to experience problems of sustainability. Moreover, even when they function, the service provided is often insufficient to allow commercial decisions to be made because of time lags between data collection and dissemination. Modern communications technologies open up the possibility for market information services to improve information delivery through SMS on cell phones and the rapid growth of FM radio stations in many developing countries offers the possibility of more localised information services. In the longer run, the internet may become an effective way of delivering information to farmers. However, problems associated with the cost and accuracy of data collection still remain to be addressed. Even when they have access to market information, farmers often require assistance in interpreting that information. For example, the market price quoted on the radio may refer to a wholesale selling price and farmers may have difficulty in translating this into a realistic price at their local assembly market. Various attempts have been made in developing countries to introduce commercial market information services but these have largely been targeted at traders, commercial farmers or exporters. It is not easy to see how small, poor farmers can generate sufficient income for a commercial service to be profitable although in India a new service introduced by Thompson Reuters was reportedly used by over 100,000 farmers in its first year of operation. Esoko in West Africa attempts to subsidize the cost of such services to farmers by charging access to a more advanced feature set of mobile-based tools to businesses.

Marketing training

Farmers frequently consider marketing as being their major problem. However, while they are able to identify such problems as poor prices, lack of transport and high post-harvest losses, they are often poorly equipped to identify potential solutions. Successful marketing requires learning new skills, new techniques and new ways of obtaining information. Extension officers working with ministries of agriculture or NGOs are often well-trained in horticultural production techniques but usually lack knowledge of marketing or post-harvest handling. Ways of helping them develop their knowledge of these areas, in order to be better able to advise farmers about market-oriented horticulture, need to be explored. While there is a range of generic guides and other training materials available from FAO and others, these should ideally be tailored to national circumstances to have maximum effect.

Enabling environments

Agricultural marketing needs to be conducted within a supportive policy, legal, institutional, macro-economic, infrastructural and bureaucratic environment. Traders and others cannot make investments in a climate of arbitrary government policy changes, such as those that restrict imports and exports or internal produce movement. Those in business cannot function if their trading activities are hampered by excessive bureaucracy. Inappropriate law can distort and reduce the efficiency of the market, increase the costs of doing business and retard the development of a competitive private sector. Poor support institutions, such as agricultural extension services, municipalities that operate markets inefficiently and export promotion bodies, can be particularly damaging. Poor roads increase the cost of doing business, reduce payments to farmers and increase prices to consumers. Finally, the ever-present problem of corruption can seriously impact on agricultural marketing efficiency in many countries by increasing the transaction costs faced by those in the marketing chain.

Recent developments

New marketing linkages between agribusiness, large retailers and farmers are gradually being developed, e.g. through contract farming, group marketing and other forms of collective action.Donors and NGOs are paying increasing attention to ways of promoting direct linkages between farmers and buyers. The growth of supermarkets, particularly in Latin America and East and South East Asia, is having a significant impact on marketing channels for horticultural, dairy and livestock products. Nevertheless, "spot" markets will continue to be important for many years, necessitating attention to infrastructure improvement such as for retail and wholesale markets.

Importance and Objectives of Agriculture Marketing

The farmer has realized the importance of adopting new techniques of production and is making efforts for more income and higher standards of living. As a consequence, the cropping pattern is no longer dictated by what he needs for his own personal consumption but what is responsive to the market in terms of prices received by him. While the trade is very organised the farmers are not Farmer is not conversant with the complexities of the marketing system which is becoming more and more complicated. The cultivator is handicapped by several disabilities as a seller. He sells his produce at an unfavorable place, time and price.

The objectives of an efficient marketing system are:

1. to enable the primary producers to get the best possible returns,
2. to provide facilities for lifting all produce, the farmers are willing, to sell at an incentive price,
3. to reduce the price difference between the primary producer and ultimate consumer, and
4. to make available all products of farm origin to consumers at reasonable price without impairing on the quality of the produce.

Facilities Needed for Agricultural Marketing

In order to have best advantage in marketing of his agricultural produce the farmer should enjoy certain basic facilities.

1. He should have proper facilities for storing his goods.
2. He should have holding capacity, in the sense, that he should be able to wait for times when he could get better prices for his produce and not dispose of his stocks immediately after the harvest when the prices are very low.
3. He should have adequate and cheap transport facilities which could enable him to take his surplus produce to the mandi rather than dispose it of in the village itself to the village money-lender-cum-merchant at low prices.
4. He should have clear information regarding the market conditions as well as about the ruling prices, otherwise may be cheated. There should be organized and regulated markets where the farmer will not be cheated by the "dalals" and "arhatiyas".
5. The number of intermediaries should be as small as possible, so that the middleman's profits are reduced. This increases! the returns to the farmers.

INADEQUACIES OF PRESENT MARKETING SYSTEM

Indian system of agricultural marketing suffers from a number of defects. As a consequence, the Indian farmer is deprived 'of a fair price for his produce. The main defects of the agricultural marketing system are discussed here.

1. Improper warehouses

There is an absence of proper ware housing facilities in the villages. Therefore, the farmer is compelled to store his products in pits, mud-vessels, "Kutcha" storehouses, etc. These unscientific methods of storing lead to considerable wastage. Approximately 1.5% of the produce gets rotten and becomes unfit for human consumption. Due to this reason supply in the village market increases substantially and the farmers are not able to get a fair price for their produce. The setting up of Central Warehousing Corporation and State Warehousing Corporation has improved the situation to some extent

2. Lack of grading and standardization

Different varieties of agricultural produce are not graded properly. The practice usually prevalent is the one known as "dara" sales wherein heap of all qualities of produce are sold in one common lot Thus the farmer producing better qualities is not assured of a better price. Hence there is no incentive to use better seeds and produce better varieties.

3. Inadequate transport facilities

Transport facilities are highly inadequate in India. Only a small number of villages are joined by railways and pucca roads to mandies. Produce has to be carried on slow moving transport vehicles like bullock carts. Obviously such means of transport cannot be used to carry produce to far-off places and the farmer has to dump his produce in nearby markets even if the price obtained in these markets is considerably low. This is even more true with perishable commodities.

4. Presence of a large number of middlemen

The chain of middlemen in the agricultural marketing is so large that the share of farmers is reduced substantially. For instance, a study of D.D. Sidhan revealed, that farmers obtain only about 53% of the price of rice, 31% being the share of middle men (the remaining 16% being the marketing cost). In the case of vegetables and fruits the share was even less, 39% in the former case and 34% in the latter. The share of middle- men in the case of vegetables was 29.5% and in the case of fruits was 46.5%. Some of the intermediaries in the agricultural marketing system are -village traders, Kutcha arhatiyas, pucca arhatiyas, brokers, wholesalers, retailers, money lenders, etc.

5. Malpractices in unregulated markets

Even now the number of unregulated markets in the country is substantially large. Arhatiyas and brokers, taking advantage of the ignorance, and illiteracy of the farmers, use unfair means to cheat them. The farmers are required to pay arhat (pledging charge) to the arhatiyas, "tulaii" (weight charge) for weighing the produce, "palledari" to unload the bullock-carts and for doing other miscellaneous types of allied works, "garda" for impurities in the produce, and a number of other undefined and unspecified charges. Another malpractice in the mandies relates to the use of wrong weights and measures in the regulated markets. Wrong weights continue to be used in some unregulated markets with the object of cheating the farmers.

6. Inadequate market information

It is often not possible for the farmers to obtain information on exact market prices in different markets. So, they accept, whatever price the traders offer to them. With a view to tackle this problem the government is using the radio and television media to broadcast market prices regularly. The news papers also keep the farmers posted with the latest changes in prices. however the price quotations are sometimes not reliable and sometimes have a great time-lag. The trader generally offers less than the price quoted by the government news media.

7. Inadequate credit facilities

Indian farmer, being poor, tries to sell off the produce immediately after the crop is harvested though prices at that time are very low. The safeguard of the farmer from such "forced sales" is to provide him credit so that he can wait for better times and better prices. Since such credit facilities are not available, the farmers are forced to take loans from money lenders, while agreeing to pledge their produce to them at less than market prices. The co-operative marketing societies have generally catered to the needs of the large farmers and the small farmers are left at the mercy of the money lenders.

Thus it is not possible to view the present agricultural marketing system in India in isolation of (and separated from) the land relations. The regulation of markets broadcasting of prices by All India Radio, improvements in transport system, etc., have undoubtedly benefited the capitalist farmers, and they are now in a better position to obtain favourable prices for their "market produce" but the above mentioned changes have not benefited the small and marginal farmers to any great extent.

CHARACTERISTICS OF AGRICULTURAL PRODUCT

Agricultural products differ in nature and contents from industrial goods in the following respects.

1. Agricultural products tend to be bulky and their weight and volume are great for their value in comparison with many industrial goods.
2. The demand on storage and transport facilities is more heavy, and more specialized in case of agricultural products than in the case of manufactured commodities.
3. Agricultural commodities are comparatively more perishable than industrial goods. Although some crops such as rice and paddy retain their quality for long time, most of the farm products are perishable and cannot remain long on the way to the final consumer without suffering loss and deterioration in quality.
4. There are certain agricultural products such as mangoes and grapes which are available only in their seasons but this condition of seasonal availability is not found in the case of industrial goods.
5. Agricultural produce is to be found scattered over a vast geographical area and as such its collection poses a serious problem. But such is not condition in the case of industrial goods.
6. There are various kinds and varieties in farm produce and so it is difficult to grade them.
7. The farmers especially in countries like India has low holding-back. Therefore he has to sell his produce immediately after the harvest at whatever price he can fetch because of his pressing needs.
8. Finally, both demand and supply of agricultural products i are inelastic. A bumper crop, without any minimum guaranteed support price from the government may spell disaster for the farmer. Similarly the farmer may not really be in a position to take advantage of shortages or deficit crop. These benefits may pass on only to the middleman.

Methods of Sale and Marketing Agencies

The marketing of agricultural produce is generally transacted in one of the following ways.

1. **Under cover or the Hatta System:** Under this system, the sale is effected by twisting or clasping the fingers of the sellers agent under cover of a cloth. The cultivator is not taken into confidence until the final bid is cleared.
2. **Open auction system:** Under this system the agent invites bids for the produce and to the highest bidder the produce is sold.
3. **Dara system:** Another related system is to keep the heaps of grains of different quantities and sell them at fiat rates without indulging in weightment etc.
4. **Moghum sale:** Under this system, sale is based on the verbal understanding between buyers and sellers and without mentioning the rate as it is understood that the buyers will pay the prevailing rate.
5. **Private agreement:** He seller may invite offers for his produce and may sell to one who might have offered the highest price for the produce.
6. **Government purchase:** The government agencies lay down fixed prices for different qualities of agriculture commodities. the sale is effected after a gradual processing for gradation and proper weightment. This practice is also followed in co-operative and regulated markets.
7. **Marketing agencies:** The various agencies engaged in the marketing of agricultural produce can be classified into two categories, viz., (i) government and quasi private agencies like the co-operative societies and (ii) private agencies. A chain of middlemen may be found operating both in Government and private agencies. The more important among these are as follows:-
   a. Merchant is the most usual purchaser of the produce, he deals in his individual capacity.
   b. Itinerant Beoparis (merchants) visit different villages, collect the produce, and take to the nearest market.
   c. Take the weighing men from the villages to the dealers in town.
   d. Agents are concerned with the assembling and distribution of agricultural produce.

EXISTING SYSTEMS OF AGRICULTURAL MARKETING IN INDIA

The existing systems of agricultural marketing in India are as briefly described here.

1. Sale to moneylenders and traders

A considerable part of the total produce is sold by the farmers to the village traders and moneylenders. According to an estimate 85% of wheat, 75% of oil seeds in U.P., 90% of jute in West Bengal and 60% of wheat, 70% of oil seeds and 35% of cotton in Punjab are sold by the farmers in the villages themselves. Often the money lenders act as a commission agent of the wholesale trader.

2. Hats and shanties

Hats are village markets often held once or twice a week, while shanties are also village markets held at longer intervals or on special occasions. The agents of the wholesale merchants, operating in different mandies also visit these markets.

The area covered by a "hat" usually varies from 5 to 10 miles. Most of "hats" are very poorly equipped, are uncovered and lack storage, drainage, and other facilities. It is important to observe that only small and marginal farmers sell their produce in such markets. The big farmers with large surplus go to the larger wholesale markets.

3. Mandies or wholesale markets

One wholesale market often serves a number of villages and is generally located in a city. In such mandies, business is carried on by arhatiyas. The farmers sell their produce to these arhatiyas with the help of brokers, who are generally the agents of arhatiyas. Because of the malpractices of these middlemen, problems of transporting the produce from villages to mandies, the small and marginal farmers are hesitant of coming to these mandies.

The arhatiyas of these mandies sell off the produce to the retail merchants. However, paddy, cotton and oilseeds are sold off to the mills for processing. The marketing system for sugarcane is different. The farmers sell their produce directly to the sugar mills.

4. Co-operative marketing

To improve the efficiency of the agricultural marketing and to save farmers from the exploitation and malpractices of middlemen, emphasis has been laid on the development of co-operative marketing societies. Such societies are formed by farmers to take advantage of collective bargaining. A marketing society collects surplus from it members and sell it in the mandi collectively. This improves the bargaining power of the members and they are able to obtain a better price for the produce. In addition to the sale of produce, these societies also serve the members in a number of other ways.

IMPROVEMENT OF AGRICULTURAL MARKETING SYSTEM

Government of India has adopted a number of measures to improve agricultural marketing, the important ones being - establishment of regulated markets, construction of warehouses, provision for grading, and standarization of produce, standarisation of weight and measures, daily broadcasting of market prices of agricultural crops on All India Radio, improvement of transport facilities, etc.

1. Marketing surveys

In the first place the government has undertaken marketing surveys of various goods and has published these surveys. These surveys have brought out the various problems connected with the marketing of goods and have made suggestions for their removal.

2. Grading and standardization

The government has done much to grade and standardize many agricultural goods. Under the Agricultural Produce (Grading and Marketing) Act the Government has set up grading stations for commodities like ghee, flour, eggs, etc. The graded goods are stamped with the seal of the Agricultural Marketing Department -AGMARK The "Agmark" goods have a wider market and command better prices. A Central Quality Control Laboratory has been set up at Nagpur and eight other regional laboratories in different parts of the country with the purpose of testing the quality and quality of agricultural products applying for the Government's "Agmark" have been created The Government is further streamlining quality control enforcement and inspection and improvement in grading.

3. Organization of regulated markets

Regulated markets have been organized with a view to protect the farmers from the malpractices of sellers and brokers. The management of such markets is done by a market committee which has nominees of the State Government, local bodies, arhatiyas, brokers and farmers. Thus all interests are represented on the committee. These committees are appointed by the Government for a specified period of time. Important functions performed by the committees can be summarized as follows.

a. fixation of charges for weighing, brokerages etc.,
b. prevention of unauthorized deductions, underhand dealings, and wrong practices by the arhatiyas,
c. enforcing the use of standardized weights,
d. providing up to date and reliable market information to the farmers, and
e. settling of disputes among the parties arising out of market operations.

The system of regulated markets has been found to be very useful in removing fraudulent practices followed by brokers and commission agents and in standardizing market practices. The committee is responsible for the licensing of brokers and weightmen. It is nested with powers to punish anyone who is found guilty of dishonest and fraudulent practices. 1t is the policy of the government to convert all markets in the country into the regulated type.

Regulated markets aim at the development of the marketing structure to have the following.

1. ensure remunerative price to the producer of agricultural commodities,
2. reduce non functional margins of the traders and commission agents, and
3. narrow down the price spread between the producer and the consumer.

To achieve these objectives, the government would go in for comprehensive and rapid expansion of regulated marketing systems. The success achieved in states like Punjab and Haryana, where regulated markets have been established in major producing areas with linked up satellite markets in the rural growth centres would be aimed at, in other areas where intensive production is taken up. The regulating marketing system has also proved a good source of generating income for the marketing boards and for use in rural infrastructure. The regulated market complex will also include facilities for grading and for monitoring of prices.

The development of regulated markets is proposed especially in areas where commercial crops like cotton, jute, tobacco and important non-traditional crops are produced and sold in weekly markets and hats. Co-operative marketing and distribution and banking will also be linked with the regulated markets. These markets will cover all the major crops. Separate market yards are proposed for livestock, fish, fruits and vegetables.

There are now over than 6,050 regulated markets with the establishment of these regulated markets. The malpractices in mandies have disappeared and the market charges have been rationalized. As much as 70% of agricultural produce is now sold in regulated markets.

In this connection, the steps taken to standardize the weight and measures in the country should be mentioned. The government has successfully replaced the different systems of weights and measures prevalent in the country with the metric system.

4. Provision of warehousing facilities

To prevent dıstress sale by the farmers, particularly the small and marginal farmers, due to prevailing low prices, rural go downs have been set up. The government has done much to provide warehousing in towns and villages.

The Central Warehousing Corporation was set up in 1957 with the purpose of constructing and running go downs and warehouses for the storage of agricultural produce. The states has set-up the State Warehousing Corporations with the same purpose. At present the Food Corporation is constructing its own network of go downs in different parts of the country. The total storage capacity in the country was 27 million tonnes at the end of the sixth plan.

5. Dissemination of market information

The government has peen giving attention to the broadcasting of market information to the farmers. Since most villages have radio sets, these broadcasts are actually heard by farmers. The newspapers also publish agricultural prices either daily or weekly accompanied by a short review of trends.

6. Directorate of marketing and inspection

The directorate was set up by the Government of India to co-ordinate the agricultural marketing of various agencies and to advise the Central and State Governments on the problems of agricultural marketing. Activities of this directorate includes the following.

a. promotion of grading and standardization of agricultural and allied commodities;
b. statutory regulation of markets and market practices;
c. training of personnel;
d. market extension;
e. market research, survey and planning and
f. administration of O>ld Storage Order, 1980 and Meat Food Products Order, 1973.

The directorate has so far formulated grade specification for 142 agricultural commodities. It enforces compulsory quality control before export on as many as 41 agricultural commodities. It is extending financial assistance to selected regulated markets for providing grading facilities for important commodities like tobacco, jute, cotton, groundnut and cashew nut at the producers level.

An allied task is the one related to marketing research and survey. This should aim at determination of best handling methods of produce to minimize losses, damage and costs, improved methods of wholesaling and retailing and planning for new marketing facilities at appropriate centres. With this aim in view, the Directorate is currently implementing two schemes.

- Market research and planning.
- Market planning and design.

Under the former scheme, the Directorate has been carrying out country-wide marketing surveys on live-stock and important agricultural and horticultural commodities to identify and study the problems of agricultural marketing. Under the latter scheme, the Directctorate has set up a Marketing Planning and Design Centre at Faridabad and a training centre and Workshop at Nagpur to study the packaging grading and marketing of selected fruits and vegetable and also advise the authorities on the designing of fruits and vegetable markets.

7. Government purchases and fixation of support prices

In addition to the measures mentioned above, the Government also announces minimum support price for various agricultural commodities from time to time in a bid to ensure fair returns to the farmers. These prices are fixed in accordance with the recommendations of the Agricultural ,Price Commission.

If the prices start falling below the declared level (say, as a result of glut in the market), the Government agencies like the Food Corporation of India intervene in the market to make direct purchase from the farmers at the support prices. These purchases are sold off by the Government at reasonable price through the public distribution system.

COOPERATIVE MARKETING

Though the above measures have improved the system of agricultural marketing to some extent, a major part of the benefits has been derived by large farmers, who have adequate marketable surplus. However, the small and marginal farmers continue to sell a major part of their produce to moneylenders to meet their credit needs and these moneylenders offer them very low prices. Therefore it is essential to form cooperatives of the small and marginal farmers to enable them to obtain fair prices for their produce. The advantages that co-operative marketing can confer on the farmer are multifarious, some of which are listed below.

1. Increases bargaining strength of the fanners

Many of the defects of the present agricultural marketing system arise because often one ignorant and illiterate farmer (as an individual) has to face well-organised mass of clever intermediaries. If the farmers join hands and for a co-operative, naturally they will be less prone to exploitation and malpractices. Instead of marketing their produce separately, they will market it together through one agency.

2. Direct dealing with final buyers

In cases, the co-operatives can altogether skip the intermediaries and enter into direct relations with the final buyers. This practice will eliminate exploiters and ensure fair prices to both the producers and the consumers.

3. Provision of credit

The marketing co-operative societies provide credit to the farmers to save them from the necessity of selling their produce immediately after harvesting. This ensures better returns to the farmers.

4. Easier and cheaper transport

Bulk transport of agricultural produce by the societies is often easier and cheaper. Sometimes the societies have their own means of transport. This further reduces cost and botheration of transporting produce to the market.

5. Storage facilities

The co-operative marketing societies generally have storage facilities. Thus the farmers can wait for better prices. Also there is no danger to their crop yield from rains, rodents and thefts.

6. Grading and standardization

This task can be done more easily for a co-operative agency than for an individual farmer. For this purpose, they can seek assistance from the government or can even evolve their own grading arrangements.

7. Market intelligence

The co-operatives can arrange to obtain data on market prices, demand and supply and other related information from the markets on a regular basis and can plan their activities accordingly.

8. Influencing marketing prices

While previously the market prices were determined by the intermediaries and merchants and the helpless farmers were mere spectators force to accept, whatever was offered to them, the co-operative societies have changed the entire complexion of the game. Wherever strong marketing co-operative are operative, they have bargained for and have achieved, better prices for their agricultural produce.

9. Provision of inputs and consumer goods

The co-operative marketing societies can easily arrange for bulk purchase of agricultural inputs, like seeds, manures fertilizers etc. and consumer goods at relatively lower price and can then distribute them to the members.

10. Processing of agricultural produce

The co-operative societies can undertake processing activities like crushing seeds, ginning 'and pressing of cotton, etc.

In addition to all these advantages, the co-operative marketing system can arouse the spirit of self-confidence and collective action in the farmers without which the programme of agricultural development, howsoever well conceived and implemented, holds no promise to success.

WAREHOUSING IN INDIA

Warehousing facilities are necessary to prevent the loss arising out of defective storage and also to equip the farmers with a convenient instrument of credit. Both the Agricultural Finance Sub-Committee (1945) and the Rural Banking Enquiry Committee (1950) emphasized the importance of ware housing as a method of promoting rural banking and finance in India.

All India Rural Credit Survey Committee (1954) recommended a three tire system of warehousing: at the national level, state and district level, village and rural level.

At present there are three main agencies in the public sector which are engaged in building large scale storage/warehousing capacity. They are: the Food Corporation of India (FCI), Central Warehousing Corporation (CWC), and State Warehousing Corporation (SWC).

FCI provides storage capacity for food grains. It has its own go downs and it also hires storage capacity from other sources such as CWC, SWC's, State Governments and private parties. In 1960-61, there were only 40 general warehouses in the countries with a total capacity of less than 0.1 million tonnes. By the end of 1988-89, the three public sector units has a storage capacity of nearly 32 million tonnes.

Besides, public sector agencies, co-operatives have also constructed warehouses in rural areas for storage of their members' produce, for stocking of fertilizers and other inputs and consumer articles. To avoid unfair competition with the go downs of the co-operative marketing societies, the state warehousing corporations do not open warehouses at any place below the sub divisional level. By 1987-88, a total storage capacity of over 10 million tonnes in the co-operative sector was available.

Kinds of warehouses

There are broadly speaking four kinds of warehouses.

They are:

a. private warehouses which are usually maintained by joint-stock companies, firms and individuals;
b. Duty-paid public ware houses which are maintained by dock authorities or port trust authorities at port
c. Bonded warehouses which are maintained either by dock authorities or by the Government and
d. licensed warehouses which are private warehouses run by co-operative societies or by private agencies, after obtaining license from the Government.

Benefits of the warehouses

The following are some of the benefits of the warehouses.

a. It gives withholding power to the agriculturist to tide over difficulties and helps them to secure better prices for their produce.
b. It gives purchasing power to traders.
c. It tends to cushion the price fluctuation and stabilize prices as it equates supply to demand.
d. It facilitates future trading.
e. It plays a very important role in implementing the agricultural price policy of the Government
f. It obviates the need for unnecessary cross-transport.
g. Huge wastages which occur owing to improper storage of agricultural produce will be minimized if warehousing develops on a large scale.
h. Warehouses render various subsidiary services, such as sorting and packing commodities for shipment, cleaning and drying goods and preparing them for the market, acting as forwarding agents for exporters of good, purchasing goods on behalf of clients, and collecting and disseminating marketing intelligence.

Progress

Storage and warehousing facilities for agricultural crops on a commercial basis are available both in the public and the private sectors. The public sector dominates and accounts for a significantly large share of the total capacity available in the economy. The main institutional agencies providing these facilities are summarized here as Central and State Warehousing Corporations, the Food Corporation of India and the cooperatives.

The Central Warehousing Corporation was set up in 1962 to do the following functions:

a. acquire and build go downs and warehouses at suitable places,
b. arrange facilities for the transport of agricultural produce .and inputs,
c. subscribe to the share capital of State Warehousing Corporations, and
d. act as agent of the Government for the purpose of purchase, sale, storage and distribution of agricultural produce and crops.

The State Warehousing Corporations have been set-up at places of state and district importance. Their functions are the same within a state as those of CWC at the national level.

IDEAL MARKETING SYSTEM

The ideal marketing system is one that maximizes the long run welfare of society. To do this, it must be physically efficient, otherwise the same output could be produced with fewer resources, and it must be electively efficient, otherwise a change in allocation could increase the total welfare and where income distribution is not a consideration.

For maximum physical efficiency, such basic physical functions as transportation, storage, and processing should be carried on in such a way so as to achieve the highest output per unit of cost incurred on them. Similarly an ideal marketing system must allocate agricultural products in time, space and form to intermediaries and consumers in such proportions and at such prices as to ensure that no other allocation would make consumers better off. To achieve this condition, prices throughout the marketing system must be efficient and must at the same time be equal to the marginal costs of production and marginal consumer utility.

The following characteristics should exist in a good marketing system.

1. There should not be any government interference in free and market transactions. The method of intervention include, restrictions on food grain movements, restrictions on the quantity to be processed, or on the construction of processing plant, price supports, rationing, price ceiling, entry of persons in the trade, etc. When these conditions are violated, the inefficiency in the market system creeps in and commodities pass into the black market. They are not then easily available at the fair prices.
2. The marketing system should operate on the basis of the independent, but systematic and orderly, decisions of the millions of the individual consumer and producers whose lives are affected by it.
3. The marketing system should be capable of developing into an intricate and far-flung marketing systems in view of the rapid development of the urban industrial economy.

4. The marketing system should bring demand and supply together and should establish an equilibrium between the two.
5. The marketing system should be able to generate employment by ensuring the development of processing industries and convincing the people to consume more processed foods, consistent with their tastes, habits and income levels.

Scientific Marketing of Farm Products

The tendency among the farmers to market their produce has been increasing. Production is complete only when the produce is marketed at a price remunerative to the farmer. Increasing specialization in production of higher marketable/ marketed surplus of the produce and alternative channels of marketing have increased the importance of the marketing activity for the farmers. However, marketing activity should be guided by certain basic principles which are briefly explained. The farmers can gain more if they follow the following principles of scientific marketing.

1. Always bring the produce for sale after cleaning it: Impurities, when present, lower the price offered by the traders-buyers in the market. The fall in price is more than the extent of impurity present in the produce would warrant. Clean produce attracts more buyers.

2. Sell different qualities of products separately: The produce of different varieties should be marketed separately. It has been observed that when different varieties of products are marketed separately, the farmers get a higher price because of the buyers preference for specific varieties.

3. Sell the produce after grading it: Graded produce is sold off quickly. The additional income generated by the adoption of grading and standardization is more than the cost incurred in the process of grading and standardization. This shows that there is an incentive for the farmers for the production of good quality products.

4. Keep abreast of market information: Price information helps him to take decisions about when and where to sell the produce, so that a better price may be obtained.

5. Carry bags/packs of standard weights: Farmers should weigh their produce and fill each bag with a fixed quantity. Majority of the farmers do not weigh their produce before taking it for sale and suffer loss by way of a possible malpractice in weighing, or they may have to make excess payments in transit (octroi, transport costs, etc.).

6. Avoid immediate post-harvest sales: The prices of the produce touch the lowest level in the peak marketing season. Farmers can get better prices by availing of warehouses facilities existing in their areas. Farmers can meet their cash needs by pledging the warehouse receipt to nationalized banks.

7. Patronize co-operative marketing societies: Farmers can get better prices by sales through a cooperative and marketing society and can avoid the possibility of being cheated. The cost of marketing particularly the transportation cost for farmers having a small quantity of marketable surplus, is minimized, for transportation is arranged co-operatively by the society and the profit earned by the society is shared among its members.

8. Sell the produce in regulated markets: The farmers should take their produce for sale to the nearly regulated markets rather than sell them in village or unregulated markets. In regulated markets marketing charges are on very few items. They get the sales slips in the regulated markets, which show the quantity of the produce marketed and the amount of charges deducted from the values of the produce. Sales slips protect farmers against the malpractices of deliberate erroneous accounting or unauthorized deductions.

COMMODITY MARKETING

The term 'commodity' is commonly used in reference to basic agricultural products that are either in their original form or have undergone only primary processing. Examples include cereals, coffee beans, sugar, palm oil, eggs, milk, fruits, vegetables, beef, cotton and rubber. A related characteristic is that the production methods, postharvest treatments and/or primary processing to which they have been subjected, have not imparted any distinguishing characteristics or attributes. Thus, within a particular grade, and with respect to a given variety, commodities coming from different suppliers, and even different countries or continents, are ready substitutes for one another. For example whilst two varieties of coffee bean, such as robusta and arabica, do have differing characteristics but two robustas, albeit from different continents, will, within the same grade band, have identical characteristics in all important respects. Agricultural commodities are generic, undifferentiated products that, since they have no other distinguishing and marketable characteristics, compete with one another on the basis of price.

Agricultural products whose production methods, postharvest treatments and/or primary processing have not imparted any distinguishing characteristics or attributes. Within a particular grade, and with respect to a given variety, commodities coming from different suppliers, and even different countries or continents, are ready substitutes for one another. This chapter describes commodity marketing processes. Five categories of commodity are discussed: grains, livestock and meat, poultry and eggs and fresh milk.

A commodity marketing system encompasses all the participants in the production, processing and marketing of an undifferentiated or unbranded farm product, including farm input suppliers, farmers, storage operators, processors, wholesalers and retailers involved in the flow of the commodity from initial inputs to the final consumer. The commodity marketing system also includes all the institutions and arrangements that effect and coordinate the successive stages of a commodity flow such as the government and its parastatals, trade associations, cooperatives, financial partners, transport groups and educational organizations related to the commodity. The commodity system framework includes the major linkages that hold the system together such as transportation, contractual coordination, vertical integration, joint ventures, tripartite marketing arrangements, and financial arrangements. The systems approach emphasises the interdependence and inter relatedness of all aspects of agribusiness, namely : from farm input supply to the growing, assembling, storage, processing, distribution and ultimate consumption of the product. The stages of commodity marketing are: assembly, transportation, storage, grading and classification, processing, packaging and distribution and retailing.

The principal participants in grain marketing are producers, marketing boards, grain elevators, brokers, millers, livestock farmers, animal feed processors, millers, other food manufacturers, grain exchanges and exporters. A common feature of grain marketing systems is the co-existence of government marketing agencies and a parallel private marketing channel. In the post-market liberalisation era many of these parastatals have either been disbanded or have been assigned a specialised role such as acting as the buyer-of-last-resort or maintaining food security reserves.

Whether storage takes place on the farm or in silos off the farm, increases in the value of products due to their time utility must be sufficient to compensate for costs at this stage, or else storage will not be profitable. Under market liberalisation, some grain marketing parastatals have suffered substantial falls in the volumes of grains which they are handling and a number of marketing parastatals now rent some of their storage capacity to farmers, grain traders and other participants in the grain marketing system. Two types of storage facility, are commonly found: the bulk storage facility where cereals are stored in concrete and/or metal bins, and the bag storage facility where the crop is stored under tarpaulin sheets. Grading usually takes place in the grain store. Grades and standards contribute to operational and pricing efficiency by providing buyers and sellers with a system of communicating price and product information. However, grading standards for domestic markets must originate from the industry itself as and when it becomes apparent that the consumer is willing to pay a higher average price for the sorted product.

Amongst the major challenges to grain marketing systems in developing countries are: rationalisation of storage facilities in terms of their location, readjustment of farmer prices to accurately reflect production and marketing costs,

Attempts by livestock farmers to adjust production to the prevailing demand are often frustrated by the lags, between changes in demand and adjustments to supply. Meat prices increase as producers withdraw animals in the short term to build up supplies in the long term. Conversely, when prices fall and farmers selling off animals to reduce production levels the act of doing so further depresses prices.

Grains for human consumption and those for animal feeds have a direct bearing upon one another. Moreover, the price of grains converted to feed will influence the costs of meat production and, therefore, the price of meat. These interrelationships are as important to government policy makers as they are to the participants in the livestock and meat marketing systems.

The major players in livestock marketing are rural traders, local cooperatives, order buyers, commissioning agents, auctions, terminal public markets and meat packers. The systems of grading beef cattle and carcasses vary by culture but may include carcass shape, fat content and meat yield. The assessment of these and other criteria is usually subjective.

The income elasticity of demand for beef in particular, and other meats in general, tends to be strongly positive in LDCs. As per capita incomes rise in developing countries, the demand for the principal livestock products meat, milk and eggs also rises. However the development of livestock industries in developing countries is sometimes constrained by the form of traditional techniques of food preparation and shopping behaviour. In many cultures it is customary to cook meat for a very long time, especially if the favoured dishes are curries and stews and in these circumstances little interest is shown in paying premium prices for more expensive cuts of meat and butchers have no incentive to add value to the product by dividing the carcass into various consumer cuts.

The poultry and egg sector is highly fragmented in most developing countries. The principal external features of an egg, which collectively determine its quality are shell texture, colour, shape and condition. The internal condition of an egg can be established using the non-destructive tests of candling. Of particular importance are the position of the yolk within the shell, yolk colour; the extent of blood spots and the translucence and firmness of the albumen and the size of the air cell inside the shell.

The weight of a poultry carcass is a primary attribute when grading the bird. The weight will vary by breed, sex, age and in accordance with the feeding regime of the bird. The eating quality of poultry meat is expressed in terms of meat tenderness, juiciness flavour and skin colour. Poultry is less of a commodity than it was at one time.

There is a close relationship between beef and milk production. In many countries, beef production is subsidiary to dairy production with sometimes over two-thirds of cattle sold for beef being animals culled from the dairy herd. The breeds and production methods which give the best milk yields rarely give corresponding results in terms of beef production; or vice versa. Milk production is seasonal with peaks just after the rains when there is lush pasture available and a progressive declines the further into the dry season. Assembly is simple and direct with milk going direct from diary farms to the processing plant. Milk collection is usually undertaken by either a marketing board or a cooperative. In general fluid milk attracts higher prices than milk destined for use in processed products. This is due to the additional costs of marketing a highly perishable product and moving and storing a very bulky commodity. The second explanatory factor is the fact that fluid milk has a lower elasticity of demand than do processed products.

Stages in a commodity marketing system

A commodity marketing system encompasses all the participants in the production, processing and marketing of an undifferentiated or unbranded farm product (such as cereals), including farm input suppliers, farmers, storage operators, processors, wholesalers and retailers involved in the flow of the commodity from initial inputs to the final consumer. The commodity marketing system also includes all the institutions and arrangements that effect and coordinate the successive stages of a commodity flow such as the government and its parastatals, trade associations, cooperatives, financial partners, transport groups and educational organisations related to the commodity. The commodity system framework includes the major linkages that hold the system together such as transportation, contractual coordination, vertical integration, joint ventures, tripartite marketing arrangements, and financial arrangements. The systems approach emphasises the interdependence and inter relatedness of all aspects of agribusiness, namely : from farm input supply to the growing, assembling, storage, processing, distribution and ultimate consumption of the product.

Grain marketing

The principal participants in grain marketing systems are producers, marketing boards, grain elevators, brokers, millers, livestock farmers, animal feed processors, millers, other food manufacturers, grain exchanges and exporters.

The physical marketing system begins with the assembling and collecting points located in the rural areas close to the producers. The next stage involves the storage areas at the national grains marketing facilities owned and operated by an appointed parastatal and/or private grain elevator; and the grain milling companies which in some countries are privately owned and in others are government enterprises. Although the size and methods of operation differ from country to country, the local assembling and collection points usually have grains brought to them either directly by the farmer-producers themselves or by rural entrepreneurs. Thus in the case of grain, the assembly and storage functions are typically combined at this marketing stage. In countries where a marketing parastatal has been given a monopoly in grain trading, private traders are sometimes authorised to buy grains from farmers (i.e. local buying agents) on behalf of the parastatal.

A common feature of grain marketing systems is the co-existence of a government marketing agency (parastatal) and a parallel private marketing channel with myriad's of private traders. Public grain marketing agencies are government appointed parastatals assigned to control or regulate the system. Prior to market liberalisation marketing parastatals were recognised, in many developing countries, as the "official channel" of the maize marketing system, and were responsible for setting the prices for major cereals (i.e. producer, into-depot, ex-depot, into-mill, ex-mill and consumer or retail). Depending on the country, these agencies usually consisted of one or more ministries related to cereals production and influencing public policies affecting food production and consumption and a parastatal established to operationalise the regulatory provisions of the public policies enacted for a given crop. In many developing countries parastatals remain important players in the grain marketing system even though their role may have changed. In the post-market liberalisation era many, but not all, of these parastatals have either been disbanded or have been assigned a specialised role such as acting as the buyer-of-last-resort or maintaining food security reserves.

The second class of actor in the commodity marketing system is private agents. These include private individuals operating in the system as petty assemblers, traders large-scale merchants, millers (both large corporations and small rural operators), brokers and retailers of grain products. The exact quantity of grain flowing through the private channel is often not known. However, most of the agents in the channel operate in rural areas and penetrate the remotest areas of the rural areas to purchase grains. A common justification for establishing parastatals is, that parastatals get to hinterlands that private operators cannot or will not reach. However, lkpi2 claims that it is private marketing agents that more often get to remote hinterlands to buy and collect maize from farmers when the government agencies fail. However, in those developing countries where national structural adjustment programme have not yet been initiated, inter-provincial grain movement controls are so strictly policed that private agents in the system cannot legally and profitably transport grain from one production zone to another. For those countries that are already adjusting their economies structurally, market liberalisation is making or will soon make much restrictions on commodity movement irrational and new form of intermediary are coming into being. Among these are commissioning agents or brokers. These entrepreneurs do not take title to the grain but take responsibility for selling the grain. They act as agents for the grain seller who may be a farmer, a grain trader or grain elevator.

Grain storage

Whether storage takes place on the farm or in silos off the farm, increases in the value of products due to their time utility must be sufficient to compensate for costs at this stage, or else storage will not be profitable. These costs will include heating, lighting, chemical treatments, store management and labour, capital investment in storage and handling equipment, interest charges and opportunity costs relating to the capital tied up in stocks. Among the less tangible costs is the risks attached to storage. These include shrinkage due to pilferage, pests, fungal growths and loss of quality due to ageing. Another risk is that demand could fall with adverse effects on prices.

Since the advent of structural adjustment programmes and market liberalisation, some grain marketing parastatals have lost their monopoly of the market and consequently the volumes of grains which they are handling has dropped substantially.

This means that they no longer require all of their storage capacity and a number of marketing parastatals now rent some of their storage capacity to farmers, grain traders and other participants in the grain marketing system.

Two types of storage facility, are commonly found, namely: the bulk storage facility where cereals are stored in concrete and/or metal bins, and the bag storage facility where the crop is stored either inside a warehouse or in the open and then covered by tarpaulin sheets. In comparative terms, the advantage of a bulk over a bag storage system are that it is more efficient because it:

- reduces congestion at the depots by not allowing for bagged maize to be dumped all over the depot yard
- reduces handling costs
- saves foreign exchange on bags, tarpaulin and fumigation, and
- lowers storage losses.

Its disadvantages are that:

- the initial invest is high, with a significant foreign currency component
- it is inflexible in terms of not being easily expandable to cope with changes in intake and off-take levels
- it relies heavily on an efficient transport system because a silo complex is only economically viable when throughout is at least 1.6 times its capacity3, and
- it needs skilled manpower to run and maintain the entire system.

On the othe®r hand, a bag system ideally overcomes the problems associated with a bulk system as enumerated above. Its main disadvantages are:

- higher quality losses in storage due to insect pests and rodents
- higher demand for foreign exchange associated with the purchase of bags
- tarpaulin and fumigation sheets.

The depot manager controls the day-to-day operations of the depot. The duties of the depot manager include accepting, checking the quality, recording the quantity and storing the produce brought in by the farmers. Each depot's record of accepted and stored produce are then sent to headquarters for necessary action. With this information to hand, inter-depot or inter-district transfers can be effected.

Grading of grains

It is important to have a grading system which accurately describes products in a uniform and meaningful manner. Grades and standards contribute to operational and pricing efficiency by providing buyers and sellers with a system of communicating price and product information. By definition, commodities are indistinguishable from one another. However, there are differences between grades and this has to be communicated to the market. By the same measure, buyers require a mechanism to signal which grades they are willing to purchase and at what premium or discount. Prices vary among the grades depending upon the relative supply of and demand for each grade. Since the value of a commodity is directly, affected by its grade, disputes can and do arise. In fact, the government may establish grading services to serve as a disinterested third party.

Grading typically occurs at the assembly stage or when a product moves into storage, during storage, or just before it leaves storage. Grading is not normally a separate marketing stage, although it has been separately identified in table 6.1 in this chapter. It is a function provided by the storage firm or the commodity merchant or the government. Prescribed procedures for grading are set forth by the trade members of commodity markets or else are stated in governments regulations. Grading may be undertaken by a member of the trade specialising in a particular commodity. Several lots of grain, oilseeds, and cotton are combined to produce a grade level required for a particular sale. This gives rise to what are known as house grades. A merchant's primary marketing advantage may be a reputation for house grades of consistent quality.

The absence of grades and standards restricts the development of effective and efficient marketing systems. For example, for some time the government of Nepal have been trying to establish an internal food marketing system. It has no nationally integrated market but due to its topography and poor communication routes there is little inter-state trade. Instead each state has closer trade relations with neighbouring Indian states. There are substantial obstacles to achieving an integrated national market as explained by Chong Yeong Lee4;

"The lack of a unified measurement system also hampers the development of marketing. In Tarai, grain is measured by weight (1 mand=37.3kg) and in the Hills and Inner Valleys it is measured by volume (1 pathi = 4.54 litres). The same unit of weight, such as "seer", is equivalent to 0.93kg of meat in the Tarai area, 0.79kg in Kathmandu and 0.25kg in Pokhara."

Clearly, the confusion which this situation gives rise to, mitigates against the successful development of a national market. Moreover, effective standardisation is basic to an efficient pricing process. Consumers use the price differentials they are willing to signal to suppliers what they want with regard to produce quantities and qualities. If produce is not in well defined units of quantity and quality then the pricing mechanism fails as a device for communicating consumer wants to suppliers. However, Dixie5 warns that any grading standards for domestic markets must originate from the industry itself as and when it becomes apparent that the consumer is willing to pay a higher average price for the sorted product. He points out that,

"Although national standards can probably be justified for export, when compulsory minimum standards are introduced for the home market it will put up prices to the consumer", and that this would,"...lower consumption and reduce the size of the local market".

Usually samples of different sizes (depending on the size of the load) are taken from each lot delivered to the depot of the buyer and these are tested for compliance with the acceptance standards. The results determine the grade into which the whole lot from which the sample came is classified to determine the price to be paid to the grower. Typical variables used in grading grain include:

- the moisture content of the grain
- the percentage of broken kernels
- degree of discolouration in the grain
- the percentage of material other than grain (MOG) in a sack or load

In Sub-Saharan Africa, for example, maize is received in 4 grades, A to D, with A being the highest grade and D the lowest. The system allows for the farmer to dispute the grade awarded to his crop. In such cases, the farmer has to submit a written request within a specific time to the depot through which his crop was delivered.

Typically, after grading is effected, the farmer is issued with a receipt showing details of the type, quantity and grade of the maize delivered. The receipt constitutes an acknowledgment by the agency that it is indebted to the farmer up to the value of the crop detailed on it. From that point on, the marketing risk attached to the crop passes from the producer to the agency/parastatal. It is relevant to point out that it usually costs the same amount to process a grain receipt for 1 tonne as it does for 1,000 tonnes.

Grain processing

Grain processing is about the most important activity from the final consumer's stand point within the marketing chain of the crop. Grain, for human consumption, is usually milled into flour or meal. Usually two types of maize flour are processed in each country, namely: the refined, white and sifted (powdery) flour produced by industrial roller mills and the unrefined and coloured maize meal produced by hammer-mills. The highly refined grade of meal is generally consumed by urban dwellers, while the hammer-mill grain meal is a low-grade "whole maize meal" favoured by rural consumers and the urban poor.

In many countries, the milling industry is highly concentrated, usually with one major company controlling over eighty percent of the maize mill trade. For example, in Kenya, maize milling is controlled by the Mercat Group of Companies (Unga Ltd, etc) (90 percent): while in Malawi, it is GRAMIL, a subsidiary of ARMARC which processes eighty three percent of the maize in that country. In Tanzania, the National Milling Corporation, (NMC) handled about eighty five percent of all maize milled before it fell into financial difficulties in the early 1990s and was allowed to go out of business; in Zimbabwe it is the National Foods which controls eighty percent of maize milling in the country until progressive liberalisation of the market increased the level of competition during the first five years of the 1990s. Grains can also undergo secondary processing and be converted into more sophisticated products such as baked foods, breakfast cereals, baby foods, cooking oils, starches, sugars etc. A considerable amount of grain is also converted to animal feed. Due to this versatility in end use, the marketing chain for grains tend to be long and complex.

Challenges for grain marketing systems

Depot network and distribution of production problems arise because crop production is rarely evenly distributed across a country. In most countries cereals production is concentrated in one or two regions of the country, and the remainder are cereals deficit areas. In these circumstances it usually becomes rational and necessary for the government agency handling grain distribution, where there is one, to construct intake depots in the producing region(s), and storage depots in the major consumption areas. But in most instances, post-independent governments in some of these countries adopted rural income policies that made it mandatory for the whole country to be provided with government agency depots within short distances of every producer. The aim was to overcome political, social and economic injustices of the past. It was a policy which was also pursued in the hope that smallholder farmers would be attracted into the cash economy. In the end, this huge investment in infrastructure was economically unsustainable and most developing countries are now locating storage facilities on the basis of economic efficiency.

Excess stock problems tend to develop because of relatively high producer prices which the governments of the countries concerned establish in order to compensate farmers whose cost for production are high. Unfortunately, these systematic increases in the producer prices usually bring them to levels that are incompatible with international supply and demand conditions. To reduce the effect of these high producer prices some governments in LDCs introduced artificial exchange rates that later did more harm than good to the export economy of these countries. In addition, local demand was low, being limited to only urban demand, for human consumption, and for livestock feed production in rural areas.

Fluctuating grain supply problem results from the crop's susceptibility to weather conditions which determine the level of harvest and, therefore, the export market supply. With each country's domestic demand often varying along with the weather, LDC countries have been known to rapidly move from a surplus to a deficit situation. In a country with high import parity and low export parity prices, the increased risk of stockpiling seems to lower the real cost of storing the excess stocks and so causes government to become indifferent toward streamlining supply with demand requirements. Low levels of intra-regional trade between developing countries is explained by subsidies paid to producers. This masks any competitive advantage in grain production that growers in one country may have over those in neighbouring countries. Where there is intra-regional trade in grains, between developing countries it is most often based on periodic shortages arising from drought civil strife. As such, export demand is usually short-lived and often financed by donor agencies, and the exporting country cannot make any long term plans to develop the trade.

ACT OF AGRICULTURAL MARKETING

The Marketing of Agricultural Products Act of 1996 was motivated more by the repeal of statutory control measures than it was by their replacement with something better. Its authors knew more about what they did not want rather than what was needed by the sector as a whole. This factor, as well as the wholesale political changes and a racially divided industry, meant that difficult questions were either glossed over or conveniently left unanswered. Consequently the present act has revealed itself to be so diluted as to be passive, with gaping holes where strategic guidelines should have been.

It is important to note however that this result is not because of the failure of its authors, as by admission they set out to draw an enabling act, which would vest the authority on the market participants themselves. They neglected to consider that pre-requisites for an act of this nature would be established and legitimate institutions (which were not there in a changing society), and democratic representation of all interests within such institutions(which was not there in a racially divided country).

- The result was a compromise act, which recognised the need for a national marketing council but failed to define its role and therefore the authority it needed to fulfil it.
- Removed statutory regulations and services but failed to set up an alternative delivery system to meet the real needs of farmers.
- Acknowledged developmental obligations but left them to market forces and the goodwill of rivals in the established sector.
- Established a the NAMC as a state organ with no real purpose by denying it any decision making or executive powers, and handing assets to the commodity trusts.
- Vested all decision-making powers in the Minister and therefore managed to create a large amount of what is basically administrative work for her office, while also concentrating state influence on the market more than ever before.

The emerging black farmers will no doubt wonder at the cruel coincidence by which government lost all its influence on the market, just as they anticipated their own place in the sun.

The process of liberalising the market also failed to provide for a tight and timely winding up of historical business. While the assets accrued under the previous act clearly were no longer the property of the producers (as with other tax provisions), a situation was created where the very same producers were called back to adjudicate over their dispersal. This has resulted in wasteful and extended conflict and litigation, while also delaying the advent of a new dispensation and the emergent farmer's turn at the trough. This dubious redistribution of what were public assets held by statutory bodies is the basis of most problems besetting the trusts. It is also one of many legal changes whose motivations seem dubious, and whose effect has been the denial of effective control over public assets for the majority.

Deregulation as a concept

There is a common but erroneous perception that agriculture has moved from a regulated to a deregulated system. Strictly speaking this is not true as there will always be a need for regulations to ensure the fair and proper conduct of participants within the industry. In reality the change has been from a state regulated to a market or self regulated system. It is this erroneous belief that the market can operate without regulations, that leads some to conclude that no dedicated authority is required to deal with marketing matters.

As long as there is a national system of regulations (whether it be state proclaimed or market determined), there will also be a need for a dedicated institute to manage and maintain it. Additionally the following characteristics inherent in markets and marketing also need continued attention:

- Since markets are all about competition, there will always be a certain amount of disputes and even conflict among the competitors. A legitimate and capable authority is therefore required to harmonise the market to keep it productive.
- Competitors by definition do not enjoy each other's successes, and unless there is a clear commercial benefit are unlikely to co-operate unless obliged to do so by an institution acting in the national rather than their selfish interests.
- The market operates according to the law of the jungle and is unlikely to adequately provide for the interests of the emerging sector, without public sector influence.
- Without statutory regulations, members need to agree on and enforce their own systems and standards without which there would be chaos. An institution is therefore required to provide both the legitimacy and support for self-regulation.
- As an industry taking its first tentative steps into free and global markets, agriculture still relies on the forced co-operation provided by statutory instruments such as levies.
- Agricultural products are related. For instance decisions taken on the marketing of animal feed impact on the beef and the poultry sectors. Government can ease, speed and even improve on the quality of interaction between the different stakeholders to the good of all.
- There are certain public good functions which the private sector is unlikely to carry out because the do not contribute directly to the producer's bottom line. labour practices, animal welfare, environmental conservation, food safety, and unfair practices are unlikely to surface unless competing producers are in dispute.
- Globalised markets require government to adopt negotiating positions, make concessions and enter agreements. The limited analytical capacity of the department and the vested interests of producers requires that government should maintain a balanced, inclusive and legitimate forum, which it can use for consultations with the industry.

- The interests of producers should be balanced against the very real concerns of both intermediate and final consumers. For instance issues around genetically modified products, the use of chemicals and the price of processing inputs should be aired and agreed on at an appropriate forum.

The role of statutory institutions

Marketing is a process that begins when investment choices are made, and carries on through production, harvesting, handling, storage, processing retailing and consumption. Marketing lends objective to all processes because there is no task in commercial agriculture that is not aimed at enhancing the profitable marketability of products. Marketing therefore has to be the underlying consideration of all agricultural development and support programmes.

There is therefore a need for an independent national statutory institution representative of all major commodities groupings within the sector to carry out those functions that the market would not be able or willing to do on its own. Such an institution would act as a clearing-house for market information and government policy, while also affording the government a balanced view and an opportunity to place social issues on the agenda of agriculture's private sector. Without such a structure, government will relinquish all official contact with and influence over the sector.

The institution needs to be independent because it will be constituted mainly of commodity representatives, and has to have some statutory powers to enforce properly reached decisions. Representatives of commodity groups would be nominated to the council, and would be joined by consumer representatives, development sector representatives, and public sector representatives appointed by the Minister. As an industry-wide council, it should have twenty or more members and be balanced in terms of private and public sector interests, producers and consumer interests, and emerging versus established sector interests.

Public and emerging sector representations are critical to the democratisation of the industry, but are hindered by the lack of an organised skills base within agriculture. Other industries face similar constrictions, but have gone around them by including African business people and experts from other disciplines. We now have high profile business people in mining and banking who were not originally in these fields. Agriculture is as much a business as any other, and the Minister should therefore cast her net wider to entrepreneurs other than farmers in her representation.

An appropriate induction programme into agriculture and agricultural issues should enhance their participatory and representational skills. Their membership of both the institution and the trusts, together with sufficient remuneration should ensure their active and effective representation of public and emerging sector interests. This policy shift is probably the best way to accelerate the participation of Africans in mainstream agriculture.

While public representatives and to some extent emerging sector participants should be remunerated to ensure quality participation, representatives of commercial commodity groupings would be entitled to compensation for expenses incurred. This is in line with the practices of all forums where delegates represent their own interests. With quality representation, a clear mandate and sufficient authority; the institution will not only assume relevance in the eyes of producers and consumers but will also have the necessary influence to give effect to government policies.

While the institution would report to the Minister because of its statutory nature, it should be empowered to implement democratically reached decisions on levies and other such instruments. Being representative it should also be charged with resolving disputes, investigations and making rulings, as well as rendering advice on local and international markets. The Minister will oversee the council's operations and hear appeals against its decisions. The present arrangement is devoid of this important distinction.

Marketing development

While the established sector was nurtured under the state protection and support offered by the control boards, presently emerging farmers have nothing to bank on. As agricultural development all over the world has always taken place and continues to take place under special circumstances, it is likely that unless a conducive environment is deliberately created for the emerging sector, then very few new farmers will break into the mainstream. At the forefront of the liberalisation of the sector even the WTO recognises this fact by providing for special and differential treatment of the developing sector.

Special marketing channels; guaranteed share of government tenders; and encouragement of the established sector to enter joint marketing deals are just a few examples of how government could help. An exciting opportunity arises from the present concerns around food security. It is imperative that government should maintain (or cause to be maintained)strategic grain reserves. If such reserves could be procured from the developing sector through specialised channels, or government started issuing part of its welfare grants as food stamps to be used at accredited developing suppliers; then a few birds could be killed with one stone.

The present act makes no special provision for the unique interests of the emerging sector, and with this omission should be seen to favour the establishment. It is important that this provision be created in the act and in the roles of both the council and the trusts. It should be a statutory provision for the use of part of the levies and trust funds towards the development of others. After-all development of presently established farmers was to a significant extent bank-rolled by their more developed colleagues and the public at large. And while some provisions are already being made in this regard, a deliberate and tighter strategy needs to be implemented by the council to this effect. Effective public and emerging sector representation will then ensure its successful implementation.

4

Consumer Demand

Throughout the world, other countries are responding to consumer concerns about food safety and the environment. The United States is considering more funding for environmental programs, such as technical assistance to farmers to learn about environmentally friendly production practices and expansion in conservation programs. Australia, New Zealand and various countries in the European Union are introducing systems of quality assurance and food safety as well as environmental farm plans, certifications and regulations to address citizens' concerns.

The Government of Canada and the provincial and territorial governments worked with the industry and interested Canadians to develop an agricultural policy for the twenty-firts century. The objective is for Canada to be the world leader in food safety, innovation and environmentally-responsible production. This policy direction recognizes the increased challenges that Canadian producers face as they work to adapt to rapid advances in technology and compete against other countries in an increasingly complex global food market.

The following is one of a series of three background briefs on key challenges that need to be addressed in building a stronger agriculture and agri-food sector in Canada:

- The effects of competition and subsidies in global markets;
- Rising consumer demands for food safety, enhanced environmental stewardship and other quality attributes; and
- The importance of skills and knowledge in an era of advancing science and technology.

SATISFYING CONSUMER DEMANDS

Consumer Demands are Changing the Face of Agriculture

As the standard of living in Canada and other developed countries has risen, per capita food consumption has stabilized and basic food needs are being met. At the same time, consumers are more discriminating about the food they buy. They want safe food, as a minimum, but they also want a greater choice of food with look, taste, nutritional value and convenience as the key factors. They also want assurances that it is produced in an environmentally-responsible manner.

Many suppliers are developing systems that demonstrate to both existing and potential customers that their products meet the specifications demanded by consumers. At the same time, these suppliers are taking advantage of changing consumer preferences to gain new markets and develop niche markets with potential price premiums, which consumers may be willing to pay.

This willingness presents opportunities in agriculture because studies show that as the average income of consumers rises, their willingness to pay for specific quality attributes also rises.

The challenge facing producers and processors is how to respond to these rising consumer demands. They must adapt if they want to capitalize on the opportunities that are available by meeting these consumer demands.

ENSURING FOOD SAFETY

Food Safety is Essential

Food safety has always been important to consumers and continues to be a basic requirement of a modern food system. Surveys show that food safety is a high priority for almost 80 per cent of Canadians. Recent high-profile food safety catastrophes include Mad Cow Disease (BSE), dioxin contamination in Europe and E-coli 0157:H7 in hamburger and unpasteurized juice here in North America.

If producers and processors cannot assure Canadians of food safety, producers and processors risk major interruptions to their business, loss of exports and a downgrading of Canada's reputation as a supplier of safe and high-quality food.

Use New Standards

These concerns are the driving force behind the use of new standards and systems to promote quality assurance and food safety. Programs based on the Hazard Analysis Critical Control Points (HACCP) and tracing systems in some parts of the food chain, particularly the processing sector, are now extending to other parts of the food chain including the farm level. Major commodity groups are developing programs.

For example, producers are tagging the ears of beef cattle to allow tracing of individual animals back to the producers. Tracking and tracing systems can also contain the cost of disease outbreaks. The recent outbreak of Foot and Mouth Disease (FMD) in the United Kingdom was very costly. It is widely acknowledged that the control of the outbreak was complicated by difficulties in tracing animal movements, due in particular to a lack of an effective identification of all farm animals.

In this case, the tracking of the movement of sheep proved difficult and in some cases impossible. Infected sheep criss-crossed the country prior to authorities realizing they had an outbreak. The Federation of Veterinary Surgeons of Europe is now calling for the identification of all farm animals with a more effective system for tracing movement internationally, within country, or between individual farms.

More Progress is Needed on the Environmental Front

Environmental concerns are also considered a high priority by 84 per cent of Canadians, who increasingly recognize the role of the environment in quality of life and human health issues. They also recognize the fundamental link that exists between agricultural production and the environment. They are placing increasing demands on farmers and processors concerning the environmental soundness of their production methods.

Producers are responding to these concerns. Many are beginning to implement environmental farm plans (19 per cent of Canadian farm operators had intentions to do so in 2001) and management practices that will ensure long-term sustainability and prosperity. Implementing these plans and practices will also help to address growing concerns about certain practices such as intensive farming operations. There has been a marked increase in the number of media reports covering the opposition to intensive farming operations in recent years.

These concerns are consistent with knowledge of the pressure points on the environment arising from farm production and practices. Agri-environmental indicators, which measure success in managing these pressure points, show mixed results, as indicated in the box below:

Agri-environmental Indicators Show Mixed Results

Risk of water contamination increased

- Percentage of farmland where nitrogen content in water has increased (more than 1 mg of nitrates per litre) between 1981 and 1996: 77% of Quebec farmland and 68% in Ontario.
- Percentage of farmland at high risk of water contamination by nitrogen in 1996: 69% in British Columbia.

Soil quality improved

- Percentage of Prairie cropland at high risk of wind erosion between 1981 and 1996 fell from 15% to 6%.
- 85% of Canadian cropland is at a tolerable risk of soil erosion by water; an improvement over the period 1981-1996.

Climate change impact

- Greenhouse gas emissions from agriculture between 1981 and 1996 increased 3.5%.

Agricultural habitat trends mixed

- Agricultural wildlife habitat increased in some regions, and either decreased or remained constant in others between 1981-1996.

Environmental planning by producers

- 35% of Ontario farmers participated in environmental farm planning workshops in 2000.
- 12% of Quebec farmers participated in agri-environmental clubs in 2000.

CHANGING DEMANDS AND THE AGRI-FOOD CHAIN

Relationships Along the Agri-food Chain are Changing

As consumers increasingly express greater concern for food safety and the environment and demand specific food quality attributes, changes are occurring throughout the agri-food chain. Major commercial buyers at the processing and retail levels, who are quite often at the forefront of changing market trends, are placing new demands on input suppliers regarding tighter specifications on attributes and methods of production.

They are establishing new marketing relationships and linkages with producers to position themselves to respond to changing consumer demands. These linkages include developing contracts with producers that combine rigid production protocols with a greater certainty on sales and guarantees on prices.

Changing Relationships in The Agri-food Chain

The drive to quality in the global food market is changing relationships across the agri-food chain. A particular example of these changes is occurring in the pork industry. A major Canadian pork processor contracts with producers for hogs with the following terms:

- The contracted producer is bound to specific production methods and must keep formal records. For example, they must vaccinate against diseases such as pneumonia as directed by veterinarians, generate ID's for individual pigs and ensure that hogs are free of drug residues.
- There are also stringent quality requirements on the final product such as acceptable fat hardness and colour score.
- In return, the producers receive a price that is comparable to those in world markets, and are assured of a more predictable cash flow. They also receive technical assistance in the form of state of the art feed and nutrition programs and animal genetics.

This type of contract also benefits the consumer through the provision of safe, nutritious and high quality food products, and provides traceablity through a formal record keeping system.

The potential for price premiums is increasingly evident in markets where producers are responding to consumer preferences for specific product attributes. In Ontario, soybean producers developed an identity preservation system that allowed them to expand sales into the lucrative Asian market for food-grade white hilum soybeans. About 30 per cent of Ontario sales are currently identity preserved. By moving quickly into niche markets, these producers have been capturing price premiums of between 10 per cent and 50 per cent.

Another example is organic food in the United States. Consumers are paying price premiums of 50 per cent for cereal crops, 60 per cent for dairy and up to 100 per cent on fruits and vegetables for what they perceive to be safer, more nutritious food.

Many countries are also considering more stringent technical standards for both domestic production and imports, including restrictions on certain varieties and increased demand for documentation of food content. Given the importance of export markets for most Canadian food products, the way Canada responds will directly affect the future growth opportunities of the agri-food sector.

AGRIBUSINESSES IN THE NEW MILLENNIUM

The last two decades of the twentieth century have witnessed unprecedented change in agriculture worldwide as a result of globalization, changes in consumer demand, the advent of new technologies, and the need to do a better job of conserving land, water and biodiversity. Consequently, agriculture in the twenty-first century is viewed as a complete system, replacing the traditional production-based concept.

In this new scenario, agribusinesses are viewed as consumer-oriented, integrated business systems which encompass primary production, processing, storage, distribution and marketing, as well as the public and private services required for them to operate competitively.

The new view of agriculture is that it is a system of value chains focused on meeting the demands and preferences of consumers, through the introduction of practices and procedures that include all activities within and outside of the production unit. In other words, all facets of agriculture are included and it is understood that the end result is not merely the production of food.

The value added of agribusinesses is much more important than the simple value of primary production. To be able to assess this value, it is important to bear in mind that five markets are involved: primary production, inputs, distribution, wholesale and retail.

This article is based on a longer text prepared by the Agribusiness Development Area for the document "Current State of Outlook for Agriculture," which IICA publishes periodically. It presents what, in our opinion, are the future challenges to agribusiness in the hemisphere.

Changes in The World Order

In May 2004, the European Union proposed eliminating subsidies on all agricultural products. It was later announced that general agreement had been reached with respect to agriculture within the WTO. A process that, it seemed, was doomed after the "Cancun failure" was thereby reactivated.

Both events augured well for the future of agribusinesses, given estimates (data summarized by Runge et al, 2003) that the overall benefits from eliminating agricultural subsidies would exceed US$32.5 billion by 2025. Despite these expectations, not all countries of the hemisphere would benefit equally and there was even the risk that some might suffer severe effects.

In order to capitalize on the new opportunities, as a group, they would have to institute structural changes involving the creation of a new public and private institutional framework; the passage of laws to facilitate and promote private enterprise, provide individuals with legal guarantees and facilitate trade and investment; the reform of laws governing land tenure and ownership; investment in infrastructure; the development of know-how in the private and social sectors; and investment in public research on agriculture and the rural milieu.

This new institutional framework would have to be consistent with commitments assumed in various international negotiations and with the national constitutional framework of each country. One common denominator in these proposed changes should be a clear message of security and confidence to the various sectors of society.

Regardless of where the different multilateral and regional agricultural negotiations go from here, the scenario at the beginning of the twenty-first century is vastly different, as reflected in the change in relative importance of the agricultural sector within the economies of the countries and in the importance of the latter in world trade.

Countries that have lower costs, that are more competitive and that are better equipped to respond to changes in demand have a great advantage over those that have been unable to institute sweeping structural reforms. Traditionally, the United States has been considered the country with the most competitive agribusinesses.

Today, However, This is Being Called into Question.

Brazil, Argentina, in the American hemisphere, and China, India and the countries of the former Soviet Union are emerging as major players in the world food trade. As a matter of fact, recent estimates suggest that, if significant measures are not taken, the United States could cease to be a net exporting country and become a net food importer.

On the world stage, the performance of China, India and Russia has a profound impact on the global economy and should therefore be carefully observed so that the countries of the American hemisphere can take strategic action to compete. China has to be watched because of its importance as both a consumer and producer, India, because of the key structural adjustments it has made and that have given it a place of prominence among developing countries, and Russia, because of its high potential to produce grains and animal products.

For the small- and medium-scale economies of the Americas, the realignment of production and trade forces involves risks as well as opportunities that will depend on their ability to produce value added goods, offer differentiated agricultural products and lower transaction costs in processes that connect producers to consumers. In the small countries, the major challenge should be to focus on reducing the steady decline in rural incomes caused by falling production and productivity levels.

In short, the countries of the Americas must develop public and private policies and rules and regulations, and build up entrepreneurship to promote and boost the development and consolidation of agribusinesses with a long-term vision. In order to become a reality, this vision must be creation of know-how.

Consumer-oriented Agribusinesses

The future of today's agribusiness is inextricably linked to trends on world markets, and their success will depend on their ability to respond to changes in such trends.

Several factors will determine the quantity, quality and type of foods that will be in demand in the future. Income level will, however, continue to be the most determining factor, followed by changes in consumption patterns.

It is well know that as income rises so does the demand for more highly processed animal products. Changes in urban populations, improvements in communications, changing perceptions on the part of consumers regarding food safety and quality, and increased awareness of the origin of and methods used to obtain foods, will continue to be constant challenges to the development of agribusiness.

To meet these challenges, technological and strategic packages will have to be developed to promote innovation that will be of benefit at the local level so that the value added is retained in the production areas and international recognition for local values is achieved.

At the same time, joint efforts are required to promote products with special characteristics, to organize in identifying market niches and implement rules and procedures to encourage the use of seals to differentiate products from one another, on the basis of origin or traditional production practices.

From the standpoint of distribution, the production-distribution-retail sales chains will continue to merge and to have increasingly less clear demarcations between them. The importance of wholesale markets around major cities will decline, and efforts will focus more on the development of the markets of origin and transparent marketing systems such as agricultural commodity exchanges and "contract agriculture."

In response to these demands, it will become necessary to find mechanisms for regional market integration; create uniform and interchangeable information systems, especially for fruits and vegetables; and establish and accept common quality standards, efficient customs procedures and financial systems capable of handling transactions in local currencies between countries at different stages of development.

The challenge will continue to be how to increase the income level for primary producers, despite rising transaction costs. This will mean that new business plans must include strategic partnerships among the different actors in the system and must seek suppliers under the best terms and conditions, including suppliers in nontraditional countries, the use of state-of-the art technologies (biotechnology), the incorporation of marketing strategies and the use of models to determine the best time to buy and sell.

The poultry industry is an excellent example of the possibilities that exist for facing the challenges posed by globalization and consumer demand through the use of these basic strategies.

Agribusinesses and Scientific Know-how

Recent developments surrounding the threat to human health associated with pathogens in foods of plant and animal origin have changed the perception of consumers and modified shopping and consumption patterns, and could impact international trade.

Since the agreements were initiated at the Uruguay Round, in the framework of the GATT, to implement sanitary and phytosanitary measures, the countries have determined that standards should be established on the basis of solid scientific principles and be applied transparently in order to avoid their being used for protectionist purposes.

The greatest challenge is therefore to make certain that agribusinesses base their development on scientific principles and that they have the institutional support and policies needed to ensure that a food or industrial agricultural product will not harm consumers when it is prepared or used properly.

This will involve the establishment of control strategies, the use of technology to reduce risk points, vertical integration of operations, certification by independent bodies and the use of equivalent risk assessment systems. The use of verification systems based on scientific principles may increase the competitiveness of local agribusinesses. However, it could have a boomerang effect by adding costs which the consumers will not want to pay.

It is important to note that the implementation of local systems can be used unilaterally to block any competition from abroad. As a result, as free trade agreements are implemented, these barriers may become more prevalent in the future.

It is therefore necessary to rely on transparent monitoring systems to be able to detect the existence of non-tariff barriers or any changes in them and provide national agribusiness with sufficient information to enable them to adjust their procedures and meet these needs.

It is likely that the use of foods prepared with genetically modified agricultural products will be the subject of debate in coming years. Advocates and critics alike agree that it will be one of the key topics at the table when discussing industry, consumers, political institutions and international organizations.

Faced with this panorama, the private sector in the Americas must take a proactive stance and keep abreast of any national and international regulations and the establishment of monitoring and risk assessment systems.

The widespread threat of terrorism, especially in the United States, has triggered responses that affect the current and future development of agribusinesses, especially the enactment of the Public Health Security & Bioterrorism Preparedness and Response Act in 2002. Immediate action is required on the part of the public and private sectors in the hemisphere to deal with the challenge of responding to these new requirements and evaluating their impacts on trade flows and the cost of merchandise.

Finally, the agribusinesses of the future will need strong innovative capability. The possibilities extend far beyond the traditional businesses to which we are accustomed or a mere improvement in value added. The production of bio-diesel or bio-fuels and/or bio-pharmaceuticals and the manufacture of products designed for specific niches make it necessary to constantly improve technological and scientific know-how.

Challenges of Future

The competitiveness of the agribusinesses of the future will be irrevocably determined by their ability to manage knowledge, and will depend on the level of professionalism of companies and of the links in the agricultural production chains. The great challenges of the future are creating the ability to understand market needs, the requirements governing trade transactions and the level of professionalism of businesses.

The ability to base business decisions on proper and timely information will be fundamental in ensuring the sustainability of agribusinesses in the twenty-first century.

These two challenges translate into other challenges. The first is the need to revise educational and training programs associated with agriculture and to adapt them to the specific needs of each country and region.

The second is the creation of infrastructure that will provide real time access to the information required to make timely business decisions, which, in turn, make it necessary to rethink programs for investment in telephony and electrification in rural areas in the hemisphere.

In short, an investment in human capital will be an essential requirement. The greatest challenge in making agribusinesses competitive will be the capacity to create new paradigms in the minds of producers and rural dwellers, so that they can adjust to change and assimilate technological changes with the required rapidity.

Staying Competitive in Agribusiness Market

In a competitive industry, management companies must continuously find ways to retain current clients while honing an edge that makes them appealing to new customers. This is especially true in these tough economic times, as many individuals and communities face financial hardship and must make tough choices about the services they purchase.

While economic turmoil and the housing slump have affected many areas of the real estate industry, management has not been hit as hard as some other areas—at least not yet. Companies are rolling with the punches and offering their new and existing clients the same high level of service, plus a few extras as well.

Competitive Business of Management

Although technology has changed the industry dramatically over the years, one thing that hasn't changed in management is the competitive nature of the industry. Thus, management firms are forced to stay ahead of the curve in order to retain their clients, keep them happy and earn new business.

"Having a full service, broad array of services and programs will certainly help firms in this market, but some boards will rely on smaller firms who charge less and have expertise in specific areas," says Andron.

"Management is an extremely competitive business, and buildings hire new management firms based on what they're paying their current firm," says Elbaz.

Usually, there's a flat fee for co-ops and condos, based on the number of units. The flat fee includes all of the basic property management functions. Because management companies are up against other companies who will undercut their prices, it behooves them to offer an array of services at an affordable rate.

That flat fee could increase, however, if a community launches a major project, such as repaving, or having their façade repointed, and the management company takes on the role of project manager.

Today's management companies still focus on the basics of managing a building, as they did years ago. But with the growth of the industry and more technology at their disposal, managers also have taken on more tasks than ever before. Co-ops, condos and HOAs can use this increase in responsibility to work with their management company, preserving the value of their community and making sure that their building are a sound investment for years to come.

Two business arrangements common within conventional agriculture are contract growing and vertical integration. These systems are largely responsible for the shift toward consolidation within agriculture and tend to be harmful to farmers, both those who have stayed outside of the industrial agriculture system and those that have become a part of it.

Contract Growing

Contract growing is used primarily in the industrial production of hogs and chickens. In this arrangement, a corporation that owns livestock contracts with farmers to raise the animals to maturity. Without a contract, a farmer must manage slaughtering, processing, and distributing as well, all while keeping sale prices low enough to compete in a corporate-controlled marketplace. From afar, it might seem like a good deal.

Here's how a contract growing system works:

- A major food corporation delivers to the farm both feed and a large number of immature animals. (In the case of chickens, the animals often arrive on the same day that they are hatched.) This allows the corporation to control the breed of the animals and the content of their feed.
- The farmer raises the animals for a pre-determined amount of time, after which the company collects, slaughters, processes and distributes them.
- The corporation returns with another round of animals and feed, and the cycle repeats.

While this system guarantees farmers a market for their animals, it also burdens them with tremendous financial risk. Farmers must pay for all feed, and absorb any financial loss from animals that die. Farmers must invest considerable capital to build structures for housing the animals, a debt they carry for years, even decades. Once they own the buildings, they must pay for maintenance, utilities and insurance costs. Farmers are also responsible for disposing of the animals' waste, which means either building receptacles for storage, selling it as fertilizer, or renting land where it can be spread.

Throughout the process, the corporation maintains full control over the animals involved. If a farmer becomes "non-competitive" (meaning he or she can't raise the animals cheaply enough), the corporation may choose not to renew the contract. But because banks will not renew loans unless a farmer has a contract, getting dumped by one corporation leaves no option but to sign up with another. Too long spent with empty barns and unpaid mortgages will drive a farmer into bankruptcy.

Because corporations rarely define exactly what "staying competitive" entails, farmers often find that the only way to survive is to cut every possible cost. Too often, that includes the costs involved with responsibly caring for the animals and appropriately handling their waste. The helpless farmer is reduced from a steward of the land to someone who manages an industrial process.

Vertical Integration

One way that major corporations maximize profits is by controlling all stages of the production and distribution processes – otherwise known as vertical integration. In addition to contracting with farms to raise their livestock, corporations will own a feed company, a farm supply company, and a processing and distribution company. This allows them to profit from every level of food production without investing in permanent assets like land or losing money to unpredictable elements such as animal mortality.

Vertical integration also gives corporations enormous control over meat and milk prices, in a fashion similar to trusts and monopolies. They buy from farmers at rock-bottom prices, then use the cheap product to drive prices down and force out competitors, which consolidates their power even more.

In theory there is nothing wrong with vertical integration, as long as corporations are operating within the boundaries of the law. After all, any business should have the right to adopt methods that maximize profits. But in agribusiness, the system has undeniable consequences. It forces farmers into debt, supports farming practices that harm the environment and abuse animals, and compromises the quality of the food we eat.

PLANNING FOR THE FUTURE

Forecasting is the process of making statements about events whose actual outcomes (typically) have not yet been observed. A commonplace example might be estimation for some variable of interest at some specified future date. Prediction is a similar, but more general term.

Both might refer to formal statistical methods employing time series, cross-sectional or longitudinal data, or alternatively to less formal judgemental methods. Usage can differ between areas of application: for example in hydrology, the terms "forecast" and "forecasting" are sometimes reserved for estimates of values at certain specific future times, while the term "prediction" is used for more general estimates, such as the number of times floods will occur over a long period.

Forecasting in Agribusiness

Appropriate forecasting is an extremely important factor for the success of any business related to the Food and Fiber Industry. Forecasting with the right techniques is needed to provide with a correct assessment for decision-making purposes. Time-Series Analysis is commonly used since most of the data sets we study in Agricultural Economics are observed over time, where the order of observation is an integral part of the analysis. So, for instance, data on prices in efficient markets are such that prices observed in close time proximity to one another are closer to each other than are prices observed at long time intervals. Data from such markets will show no particular devotion to their historical mean. Many standard (non-time series) statistical and econometric methods rely on an informative mean for their desirable properties. Failure to correctly account for the time series properties of one's data may result in the fitting of spurious relations between or among economic data sets.

Risk and uncertainty are central to forecasting and prediction; it is generally considered good practice to indicate the degree of uncertainty attaching to forecasts. The process of climate change and increasing energy prices has led to the usage of Egain Forecasting of buildings. The method uses Forecasting to reduce the energy needed to heat the building, thus reducing the emission of greenhouse gases.

Practice of Customer Demand Planning

Forecasting is used in the practice of Customer Demand Planning in every day business forecasting for manufacturing companies. The discipline of demand planning, also sometimes referred to as supply chain forecasting, embraces both statistical forecasting and a consensus process. An important, albeit often ignored aspect of forecasting, is the relationship it holds with planning. Forecasting can be described as predicting what the future will look like, whereas planning predicts what the future should look like.

There is no single right forecasting method to use. Selection of a method should be based on your objectives and your conditions (data etc.). A good place to find a method, is by visiting a selection tree. An example of a selection tree can be found here.

While the term "forecasting" may appear to be rather technical, planning for the future is a critical aspect of managing any organization—business, nonprofit, or other.

In fact, the long-term success of any organization is closely tied to how well the management of the organization is able to foresee its future and to develop appropriate strategies to deal with likely future scenarios. Intuition, good judgment, and an awareness of how well the economy is doing may give the manager of a business firm a rough idea (or "feeling") of what is likely to happen in the future.

Nevertheless, it is not easy to convert a feeling about the future into a precise and useful number, such as next year's sales volume or the raw material cost per unit of output. Forecasting methods can help estimate many such future aspects of a business operation.

Suppose that a forecast expert has been asked to provide estimates of the sales volume for a particular product for the next four quarters. One can easily see that a number of other decisions will be affected by the forecasts or estimates of sales volumes provided by the forecaster. Clearly, production schedules, raw material purchasing plans, policies regarding inventories, and sales quotas will be affected by such forecasts. As a result, poor forecasts or estimates may lead to poor planning and thus result in increased costs to the business.

How should one go about preparing the quarterly sales volume forecasts? One will certainly want to review the actual sales data for the product in question for past periods. Suppose that the forecaster has access to actual sales data for each quarter over the 25 year period the firm has been in business. Using these historical data, the forecaster can identify the general level of sales. He or she can also determine whether there is a pattern or trend, such as an increase or decrease in sales volume over time. A further review of the data may reveal some type of seasonal pattern, such as peak sales occurring before a holiday.

Thus by reviewing historical data over time, the forecaster can often develop a good understanding of the previous pattern of sales. Understanding such a pattern can often lead to better forecasts of future sales of the product.

In addition, if the forecaster is able to identify the factors that influence sales, historical data on these factors (or variables) can also be used to generate forecasts of future sales volumes.

FORECASTING METHODS

All forecasting methods can be divided into two broad categories: qualitative and quantitative. Many forecasting techniques use past or historical data in the form of time series. A time series is simply a set of observations measured at successive points in time or over successive periods of time. Forecasts essentially provide future values of the time series on a specific variable such as sales volume. Division of forecasting methods into qualitative and quantitative categories is based on the availability of historical time series data.

QUALITATIVE FORECASTING METHODS

Qualitative forecasting techniques generally employ the judgment of experts in the appropriate field to generate forecasts. A key advantage of these procedures is that they can be applied in situations where historical data are simply not available. Moreover, even when historical data are available, significant changes in environmental conditions affecting the relevant time series may make the use of past data irrelevant and questionable in forecasting future values of the time series. Consider, for example, that historical data on gasoline sales are available.

If the government then implemented a gasoline rationing program, changing the way gasoline is sold, one would have to question the validity of a gasoline sales forecast based on the past data. Qualitative forecasting methods offer a way to generate forecasts in such cases. Three important qualitative forecasting methods are: the Delphi technique, scenario writing, and the subject approach.

Delphi Technique

In the Delphi technique, an attempt is made to develop forecasts through "group consensus." Usually, a panel of experts is asked to respond to a series of questionnaires. The experts, physically separated from and unknown to each other, are asked to respond to an initial questionnaire (a set of questions). Then, a second questionnaire is prepared incorporating information and opinions of the whole group. Each expert is asked to reconsider and to revise his or her initial response to the questions.

This process is continued until some degree of consensus among experts is reached. It should be noted that the objective of the Delphi technique is not to produce a single answer at the end. Instead, it attempts to produce a relatively narrow spread of opinions—the range in which opinions of the majority of experts lie.

Scenario Writing

Under this approach, the forecaster starts with different sets of assumptions. For each set of assumptions, a likely scenario of the business outcome is charted out.

Thus, the forecaster would be able to generate many different future scenarios (corresponding to the different sets of assumptions). The decision-maker or businessperson is presented with the different scenarios, and has to decide which scenario is most likely to prevail.

SUBJECTIVE APPROACH

The subjective approach allows individuals participating in the forecasting decision to arrive at a forecast based on their subjective feelings and ideas. This approach is based on the premise that a human mind can arrive at a decision based on factors that are often very difficult to quantify.

"Brainstorming sessions" are frequently used as a way to develop new ideas or to solve complex problems.

In loosely organized sessions, participants feel free from peer pressure and, more importantly, can express their views and ideas without fear of criticism. Many corporations in the United States have started to increasingly use the subjective approach.

QUANTITATIVE FORECASTING

Quantitative forecasting methods are used when historical data on variables of interest are available—these methods are based on an analysis of historical data concerning the time series of the specific variable of interest and possibly other related time series. There are two major categories of quantitative forecasting methods.

The first type uses the past trend of a particular variable to base the future forecast of the variable. As this category of forecasting methods simply uses time series on past data of the variable that is being forecasted, these techniques are called time series methods.

The second category of quantitative forecasting techniques also uses historical data. But in forecasting future values of a variable, the forecaster examines the cause-and-effect relationships of the variable with other relevant variables such as the level of consumer confidence, changes in consumers' disposable incomes, the interest rate at which consumers can finance their spending through borrowing, and the state of the economy represented by such variables as the unemployment rate.

Thus, this category of forecasting techniques uses past time series on many relevant variables to produce the forecast for the variable of interest. Forecasting techniques falling under this category are called causal methods, as the basis of such forecasting is the cause-and-effect relationship between the variable forecasted and other time series selected to help in generating the forecasts.

UNDERSTANDING OF BEHAVIOUR OF TIME SERIES

Before discussing time series methods, it is helpful to understand the behavior of time series in general terms. Time series are comprised of four separate components: trend component, cyclical component, seasonal component, and irregular component. These four components are viewed as providing specific values for the time series when combined.

In a time series, measurements are taken at successive points or over successive periods. The measurements may be taken every hour, day, week, month, or year, or at any other regular or irregular interval. While most time series data generally display some random fluctuations, the time series may still show gradual shifts to relatively higher or lower values over an extended period. The gradual shifting of the time series is often referred to by professional forecasters as the trend in the time series. A trend emerges due to one or more long-term factors, such as changes in population size, changes in the demographic characteristics of population, and changes in tastes and preferences of consumers. For example, manufacturers of automobiles in the United States may see that there are substantial variations in automobile sales from one month to the next. But, in reviewing auto sales over the past fifteen to twenty years, the automobile manufacturers may discover a gradual increase in annual sales volume.

In this case, the trend for auto sales is increasing over time. In another example, the trend may be decreasing over time. Professional forecasters often describe an increasing trend by an upward sloping straight line and a decreasing trend by a downward sloping straight line. Using a straight line to represent a trend, however, is a mere simplification—in many situations, nonlinear trends may more accurately represent the true trend in the time series.

Although, a time series may often exhibit a trend over a long period, it may also display alternating sequences of points that lie above and below the trend line. Any recurring sequence of points above and below the trend line that last more than a year is considered to constitute the cyclical component of the time series—that is, these observations in the time series deviate from the trend due to cyclical fluctuations (fluctuations that repeat at intervals of more than one year). The time series of the aggregate output in the economy (called the real gross domestic product) provides a good example of a time series that displays cyclical behavior. While the trend line for gross domestic product (GDP) is upward sloping, the output growth displays a cyclical behavior around the trend line. This cyclical behavior of GDP has been dubbed business cycles by economists.

The seasonal component is similar to the cyclical component in that they both refer to some regular fluctuations in a time series. There is one key difference, however. While cyclical components of a time series are identified by analyzing multiyear movements in historical data, seasonal components capture the regular pattern of variability in the time series within one-year periods. Many economic variables display seasonal patterns. For example, manufacturers of swimming pools experience low sales in fall and winter months, but they witness peak sales of swimming pools during spring and summer months.

Manufacturers of snow removal equipment, on the other hand, experience the exactly opposite yearly sales pattern. The component of the time series that captures the variability in the data due to seasonal fluctuations is called the seasonal component.

The irregular component of the time series represents the residual left in an observation of the time series once the effects due to trend, cyclical, and seasonal components are extracted. Trend, cyclical, and seasonal components are considered to account for systematic variations in the time series. 'h e irregular component thus accounts for the random variability in the time series. The random variations in the time series are, in turn, caused by short-term, unanticipated and nonrecurring factors that affect the time series. The irregular component of the time series, by nature, cannot be predicted in advance.

5

Agriculture Income, Output, Price Indices

One of the principal objectives of the common agricultural policy (CAP) is to provide farmers with a reasonable standard of living. Although this concept is not defined explicitly within the CAP, a range of indicators including income development from farming activities may be used to determine the progress being made towards this objective.

According to second estimates, real agricultural income per worker in the European Union (EU) has decreased by 16.6% in 2009, following a decrease by 1.8% in 2008. This decline results from a reduction in real agricultural income (-13.6%), together with a fall in agricultural labour input (-2.3%). These estimates for the EU are based on data supplied by the national authorities in the Member States.

Main statistical findings

Second estimates for 2010: down by 11,6% on the previous year. The decrease in EU real agricultural income in 2009 is mainly the result of a sharp fall in the value of agricultural output at producer prices in real terms (-10.5%), while input costs in real terms decreased (-9.2%). The small fall in the real value of subsidies net of taxes (-0.5%) and the slight decrease in depreciation in real terms (-0.2%) have a marginal impact.

Real agricultural income per worker in 2009 is estimated to have fallen in 21 Member States and to have risen in six. The largest falls are expected in Hungary (-32.2%), Luxembourg (-25.2%), Ireland (-23.6%), Germany (-21.0%), Italy (-20.6%), Austria (-19.4%) and France (-19.0%), and the highest rises in Malta (+7.8%), Denmark, (+4.3%) and Finland (+2.6%).

In 2009, the value of EU agricultural output at producer prices is estimated to have decreased by 10.7%, mainly due to a decrease in the value of both crop production (-13.0%) and animal production (-9.3%) in real terms.

- Declining crop prices: In crop production, the decrease in value is almost entirely due to a decline in prices (-12.8%), while the volume is expected to fall only slightly (-0.3%). Prices are falling for all groups of crops, except plants and flowers (+1.5%), most sharply forcereals (-27.1%), oilseeds (-24.8%), olive oil (-16.3%) and fruits (-15.9%). A rise in volume is seen for oilseeds (+10.8%), while olive oil (-9.3%) and cereals (-5.7%) are declining.
- Falling prices for animal production: The decrease in the value of animal production in 2009 is the result of a clear fall in producer prices (-8.3%) and a slight decline in volume (-1.0%). Prices are falling for the three largest animal

products: milk (-20.6%), pigs (-3.8%) and cattle (-1.2%). The volume is nearly stable for milk (-0.5%), unchanged for pigs and decreases for cattle (-3.0%).

- Decreasing input costs: EU agricultural input costs (intermediate consumption) are expected to fall by 9.2% in real terms, due to decreases in both volume (-2.7%) and prices (-6.7%). Lower input volumes are estimated for almost all categories of input, particularly for fertilizers (-14.6%). The fall in input prices is driven by sharp falls in feeding stuff (-16.7%) and energy (-14.5%).

Between 2005 and 2009, EU-27 real income per worker is estimated to have decreased by 0.3%. On average, agricultural labour input in the EU-27 has fallen by 11.5% since 2005.

DATA SOURCES AND AVAILABILITY

Estimated results for agricultural income and labour input are submitted by Member States for the year in question by the end of November the same year and by the end of February the following year. Final data are submitted in September, 9 months after the end of the year.

Output and subsidies

In value terms, the EU-27's crop output grew 9.1 % in 2011 to EUR 203 330 million and animal outputincreased by 9.9 % to EUR 154 057 million; these increases in gross output were offset by a sharp increase in the value of intermediate consumption of goods and services at basic prices (up 11.1 %). As a result, the agricultural sector generated EUR 148 556 million of gross value added at producer prices in 2011, which represented a 6.0 % increase in relation to the previous year.

Value added at producer prices reflects the value of output before taking account of subsidies and taxes on products and production. Value added at basic prices adds in any subsidies (less taxes) on products: in 2011 the EU-27's agricultural sector recorded value added at basic prices that was EUR 153 000 million, 3.0 % higher than the valuation at producer prices. Value added at factor cost also adds in subsidies (less taxes) on production: in 2011 this was valued at EUR 198 932 million, a further 30.0 % higher than the valuation at basic prices. The various valuations of value added reflect the relative importance of product and production subsidies and to a lesser extent taxes. In 2011 total subsidies (before the deduction of taxes) on products for the agricultural sector were EUR 4 709 million, while subsidies on production were more than ten times this level, EUR 50 821 million; after deducting taxes these levels were EUR 4 445 million for products and EUR 45 932 million for production.

Types of agricultural subsidy have changed over time through successive reforms of the CAP: the shift from product to production subsidies was mainly implemented in 2005 and 2006. Whereas production subsidies were ten times as high as product subsidies in 2011, in 2001 product subsidies had been nearly twice as high as production subsidies. In total these two types of subsidy were equivalent to an extra 37.4 % of value added at producer prices in 2011 in the EU-27. The net impact of subsidies less taxes in 2011 was to add an extra 33.9 % to value added at producer prices; for comparison, in 2001 the increase due to subsidies less taxes was 25.8 %.

Price indices

Changes in the value of agricultural output comprise a volume and price component: one important strand of recent changes in agricultural policy has been to move away from price support mechanisms, so that prices more accurately reflect market forces and changes in supply and demand. During the period 2006 to 2011 there were considerable differences between the EU Member States in the development ofdeflated agricultural output prices; such deflated prices show the extent to which agricultural prices have changed compared with consumer prices. Deflated output prices rose in 22 of the EU Member States, the largest increases being recorded for the United Kingdom (average growth of 7.0 % per annum) and Romania (4.3 % per annum), while reductions were posted in five of the EU Member States, the largest being in Belgium (-1.7 % per annum) and Cyprus (-1.4 % per annum).

The development of deflated agricultural input prices showed a broadly similar picture, as prices rose in 24 of the EU Member States, with the United Kingdom again recording the highest input price increases. Cyprus figured at the other end of the range, in this case with by far the largest decrease in input prices (-2.2 % per annum). For 14 of the EU Member States the change in output prices was greater than the change in input prices, with the largest (percentage point) differences observed for Romania and the United Kingdom. Among the 13 EU Member States where input price changes exceeded the change in output prices the greatest differences were observed for Slovenia, Spain, Belgium and Portugal.

There was an overall 28.3 % increase in EU-27 output prices for agricultural goods between 2005 and 2011, which could be broken down for the main components into increases of 33.7 % for crop output and 22.8 % for animal output. The upward development of output prices between 2005 and 2011 did not follow a smooth pattern, as there was a considerable reduction in prices between 2008 and 2009 when the price of agricultural goods fell by 11.5 %, which could be largely attributed to falling prices for see Table 3. Equally, the 9.6 % increase in output prices for agricultural goods between 2010 and 2011 was led by a 34.7 % increase for cereals, while fruits (-0.3 %) and vegetables and horticultural products (-6.8 %) were the only product categories to record falling output prices in 2011.

Income: Indicator A

The real net value added at factor cost of agricultural activity per unit of labour (expressed in annual work units), also known as the agricultural income indicator A, increased by 7.1 % in the EU-27 from 2010 to 2011. There were stark contrasts among the EU Member States in the development of this indicator, with income rising at a rapid pace in Romania (56.8 %) and Hungary (49.2 %), in contrast to falling income in Belgium (-22.5 %), Malta (-14.0 %), Portugal (-10.3 %) and the Netherlands (-8.9 %) and to a lesser extent in Finland, Greece, France and Spain.

Data sources and availability

Economic accounts for agriculture (EAA) provide an insight into:

- the economic viability of agriculture;
- agriculture's contribution to each EU Member State's wealth;
- the structure and composition of agricultural production and inputs;
- the remuneration of factors of production;
- relationships between prices and quantities of both inputs and outputs.

These accounts comprise a production account, a generation of income account, an entrepreneurial income account and some elements of a capital account. For the production items, EU Member States transmit toEurostat values at basic prices, as well as their components (values at producer prices, subsidies on products, and taxes on products). The data for the production account and for gross fixed capital formation are transmitted in both current prices and the prices of the previous year.

The output of agricultural activity includes output sold (including trade in agricultural goods and services between agricultural units), changes in stocks, output for own final use (own final consumption and own-account gross fixed capital formation), output produced for further processing by agricultural producers, as well as intra-unit consumption of livestock feed products. The output of the agricultural sector is made up of the sum of the output of agricultural products and of the goods and services produced in inseparable non-agricultural secondary activities; animal and crop output are the main product categories of agricultural output.

Gross value added equals the value of output less the value of intermediate consumption and is shown in producer prices (the producer price excludes subsidies less taxes on products). Intermediate consumption represents the value of all goods and services used as inputs in the production process, excluding fixed assets whose consumption is recorded as fixed capital consumption. The EU Member States transmit information on intermediate consumption to Eurostat using values at purchaser prices (basic prices).

Eurostat also collects annual agricultural prices (in principle net of VAT) to compare agricultural price levels between EU Member States and to study sales channels. Quarterly and annual price indices for agricultural products and the means of agricultural production, on the other hand, are used principally to analyse price developments and their effect on agricultural income. Agricultural price indices are obtained by a base-weighted Laspeyres calculation (2005=100), and are expressed in nominal terms or as deflated indices based on the use of an implicit consumer prices (HICP) deflator.

Agricultural income indicators are presented in the form of:

- an index of real income of factors in agricultural activity per annual work unit (indicator A);

- an index of real net agricultural entrepreneurial income, per unpaid annual work unit (indicator B);
- and the net entrepreneurial income of agriculture (indicator C).

The information presented in this article for indicator A relates to prices of the previous year - in other words, an example value of 107.1 represents a 7.1 % increase (based on the prices of the previous year).

Context

Significant reforms of the common agricultural policy (CAP) have taken place in recent years, most notably in 2003 and 2008, with the aim of making the agricultural sector more market-oriented. The 2003 reform introduced a new system of direct payments, known as the single payment scheme, under which aid is no longer linked to the level of production (decoupling); this single payment scheme aims to guarantee farmers more stable incomes. Farmers can decide what to produce in the knowledge that they will receive the same amount of aid, allowing them to adjust production to suit demand. In 2008 further changes were made to the CAP, building on the reform package from 2003, such that all aid to the agricultural sector should have been decoupled by 2012.

The Europe 2020 strategy offers a new perspective on economic, social, environmental, climate-related and technological challenges and future agricultural reform is likely to be made in relation to the goals of developing intelligent, sustainable and inclusive growth, while taking account of the wealth and diversity of the agricultural sector within the EU Member States. As part of this process, the European Commission launched a public debate on the future of the CAP during 2010. Its outcome, coupled with input from the European Council and Parliament led the Commission to present a Communication in November 2010, titled 'The CAP towards 2020: meeting the food, natural resources and territorial challenges of the future' (COM(2010) 672 final). This was followed, in October 2011, by a set of legal proposals concerning the future of the CAP: it is planned that this revised agricultural policy will be in place by January 2014 - for more information, see the introductory article for agriculture, forestry and fisheries.

AGRICULTURAL INCOME PER ANNUAL WORK UNIT

According to the second estimates of the economic accounts for agriculture (EAA) submitted to Eurostat by the European Union (EU) Member States at the end of January 2013, agricultural income per annual work unit (Indicator A) at EU-27 level is estimated to have remained stable in 2012 compared with 2011. However, agricultural labour input decreased slightly (- 0.2 %) in EU-27 compared with the previous year.

Main statistical findings

Agricultural income per annual work unit remained stable at the EU-27 level in 2012, following an increase of 8.3 % in 2011 and 17.1 % in 2010.

This latest development was driven by an increase of 3.6 % in agricultural income per work unit in EU-15, which was counterbalanced by a decrease in those Member States which joined the EU in 2004 and 2007, (-11.9 %). The decrease in Indicator A for Member States which joined the EU in 2004 and 2007 was linked to the decrease in factor income (-11.3 %). Nonetheless, the agricultural labour input registered a small increase of 0.7 % in 2012.

On the other hand, the factor income in EU-15 increased by 2.4 % in 2012 compared with the previous year. This led to an increase of 3.6 % in the agricultural income, which when divided by the agricultural labour input, resulted in a decrease in Indicator A (-1.1 %).

For EU-27, the index (2005=100) of Indicator A reached 128.5 % in 2012, the same as in 2011. This increase since 2005 is attributable to the positive developments in EU-15 and Member States which joined the EU in 2004 and 2007, which recorded rises of 17.9 % and 50.1 % respectively, when compared with 2005.

Agricultural income per work unit developed differently across the Member States of the EU-27. In 2012, the largest increases were seen in Belgium (+27.9 %), Latvia (+21 %) and Lithuania (+17.6 %). The Netherlands, Germany and Estonia also showed increases of over 14 %.

The steepest decreases, on the other hand, were seen in Romania (-27.1 %), Poland (-13.7 %) and Slovenia (-12.2 %). Hungary, the United Kingdom, Ireland and Austria also recorded marked decreases (steeper than -7.5 %).

In 2012, almost 83 % of total factor income in EU-27 was generated in EU-15, by 50 % of the total agricultural labour input (in AWU).

The trends in the two EFTA countries for which data were available, were very different in 2012 compared with 2011. Indicator A rose in Norway (+10.7 %) while it decreased in Switzerland (-1.1 %).

Croatia, the new EU Member State as of July 2013, observed a decrease in agricultural income per work unit, Indicator A (-10.8 %).

Main factors in income development for 2012

Indicator A combines the development in net value added at factor cost (factor income) and the trend in agricultural labour input .

In EU-27, there was a slight decrease (-0.2 %) in factor income for 2012 compared with 2011 in real terms (which represents an increase of 1.2 % in nominal terms), while agricultural labour input also decreased slightly (-0.2 %).

Factor income development in 2012 was mainly attributable to small increases in the output values of bothcrop production (+0.9 %) and animal production (+3.2 %). The output value of agricultural services registered a slight increase in 2012 compared with 2011 (+0.7 %). On the other hand, the output value of inseparable non-agricultural secondary activities dropped (-1.2 %).

Even though the overall output of the agriculture industry grew by 1.8 %, according to the second estimates, this was counterbalanced by an identical increase of 1.8 % in the value of intermediate consumption. This led to factor income remaining unchanged.

The stable level of agricultural income per labour unit in 2012 compared with 2011 results from a combination of small decreases in the volume of crop and animal production compensated by slight increases in prices.

Higher prices in agriculture (+6.6 %) led to an increase of 1.8 % in the value of total agricultural output in basic prices. On the other hand, the volume of agricultural output decreased (-3.0 %).

For intermediate consumption, the increase in value (+1.8 %) also resulted from a rise in prices (+4.9 %) and a decrease in volume (-1.5 %). As a result of both of these trends, i.e. the developments in output and intermediate consumption, the gross value added in agriculture in 2012 showed the same trend (+1.8 %) as the value of total agricultural output at basic prices.

The gross value added at basic prices accounted for 39.3 % of the total output value. Overall, subsidies were worth more than EUR 55 thousand million in 2012, accounting for around 38 % of the factor income.

A closer look at output

The value of the output of agricultural goods at basic prices increased by 1.9 % in 2012 compared with 2011. This small increase was driven by a decrease in volume (-3.2 %) and an increase of 6.9 % in the output prices of agricultural goods. The value of crop output at basic prices, which accounted for 52.2 % of the total output, remained at almost the same level. On the other hand, animal output, at around 40 % of total output, recorded a rise of 3.2 %. Agricultural services increased slightly (+0.7 %) while secondary activities decreased (-1.2 %).

Crop output

The value of crop output at basic prices increased slightly in 2012, compared with 2011 (+0.9 %), while animal output showed a larger increase (+3.2 %). The reason for the stability of crop output was a combination of a decrease in volume (-5.3 %)and the increase in producer prices (+6.9 %). Price increases were observed for almost all main agricultural crops. The largest price increases in 2012 were observed for cereals (+10.0 %), potatoes (+9.2 %), oil seeds and wine (+8.8 %). A small decrease in the producer price was observed only for sugar beet (-0.2 %) in 2012 compared with 2011. On the other hand, the changes in volume were negative for most agricultural crops (wine, -15.7 %; potatoes, -14.1 %) while oil seeds, fruits and cereals registered a volume decrease steeper than -7 %.

Ten Member States reported declines in total crop output in 2012, which were most marked in Romania (-28.9 %), Slovakia (-11.0 %), Malta (-8.5 %) and Slovenia (-7.9 %). The largest increases in the value of crop output were observed in Latvia and Luxembourg (+31.4 %), Estonia (+16.4 %), Lithuania (+15.1 %) and Belgium (+13.5 %).

The value of crop output at basic prices in EU-15 increased by 3.2 % in 2012, while in Member States which joined the EU in 2004 and 2007, it decreased (-8.6 %). Basic prices in real terms rose by 6.5 % in EU-27, as a result of the increase of 7.1 % in EU-15 and 3.5 % in the other Member States.

Animal output

The rise in the value of animal output recorded in 2012 (+3.2 %) was mainly due to higher producer prices (+3.8 %) while the volume of animal output declined slightly (-0.5 %). At the EU-27 level, prices rose for most of the groups included in animal output. For animal products, the value of milk, which accounted for 13.6 % of the overall output in 2012, decreased (-4.4 %) compared with 2011, mainly due to the decrease in the price of milk (-4.7 %). On the other hand, the volume remained almost at the same level (+0.3 %) compared with 2011. As regards egg production, the large increase in the producer price (+30.6 %) brought about an increase of 28.5 % in value when compared with the previous year.

For animals (meat and livestock), the value in basic prices was up 5.5 %. Higher producer prices were the main reason for this in almost all animal groups: equines (+13.1 %), pigs (+9.7 %), cattle (+8.4 %) and poultry (+2.3 %). The only (slight) decrease in the price development was observed for sheep and goats (-1.0%). As regards volume, all categories (except poultry, +2.6 % and milk, +0.3 %) registered a drop compared with 2011: sheep and goats (-2.2 %); cattle (-2.1 %) and pigs (-1.8 %).

20 out of the 27 EU Member States reported higher values for animal output in 2012 than in 2011, with the most noticeable increases observed in Spain (+9.8 %), Hungary (+6.8 %) and Romania (+5.5 %). Member States such as Sweden (-5.9 %), Slovakia (-4.8 %) and Lithuania and Luxembourg (-2.0 %), on the other hand, reported lower output values in real terms than in the previous year.

A medium-term analysis (2005-2012)

In order to place the most recent income developments in a medium-term perspective, Indicator A is expressed in index form, with the year 2005 chosen as reference year. According to the trend of Indicator A, Member States can be divided into two groups.

The first group includes those countries for which agricultural income in 2012 was higher than in 2005. This group comprises twenty-two Member States. In fifteen of these, the level of agricultural income per work unit reached in 2012 was more than 20 index points higher than in 2005. The most marked trends were reported by Estonia (+119 %), Slovakia (+95.9 %) and Lithuania (+81.6 %).

There were five Member States which had a negative trend for Indicator A in 2012 compared with 2005: Cyprus, Luxembourg, Malta, Ireland and Italy.

Value of intermediate consumption was higher in 2012 than 2011

In 2012, at the EU level, the value of intermediate consumption increased by 1.8 % in real terms compared with 2011, due to a combination of higher prices (+3.3 %) partially offset by a decrease in volume (-1.5 %). As regards the components of intermediate consumption, values for most items rose moderately: financial intermediation services indirectly measured (+4.5 %); feedingstuffs (+3.3 %); energy and lubricants (+3.2 %) and plant protection products (+1.9 %). The only declines compared with the previous year were observed in the values of seeds and planting stocks (-1.9 %) and veterinary expenses (-1.1 %).

The value of intermediate consumption rose in twenty-two EU Member States in 2012, compared with the previous year. The most marked increases were recorded in Latvia (+12.6 %), Luxembourg (+8.6 %) and Estonia (+8.0 %). Out of the five countries which reported a decrease in 2012, the steepest declines were observed in Romania (-17.4%), Slovakia (-5.6%) and Malta (-4.9%).

Energy prices rose in all Member States in 2012, compared with 2011 (except Denmark, where a slight decrease, -1.3 %, was recorded). The increase of energy prices ranged from 1.1 % in the United Kingdom to 11.1 % in France. The largest increases in the prices of fertilisers and soil improvers were recorded in Bulgaria (+14.9 %), France (+12.6 %), Greece (+11.8 %) and Romania (+10.1 %).

Agricultural labour input remained at the same level in 2012

In 2012, the agricultural labour input decreased slightly in thirteen Member States compared with the previous year; it remained almost stable in eight and increased in six. The steepest declines were seen in Slovakia (-5.7 %), Sweden (-4.7 %) and Finland and the Netherlands (-3.3 %).

For EU-27, agricultural labour input remained at the same level in 2012 compared with 2011; however, it decreased markedly (-19.8 %) compared with 2005 and even more so (-32.2 %) relative to 2000. The labour input used in agriculture in EU-15 decreased in 2012 (-1.1%) compared with 2011, (2012 vs. 2005, -13.9%; 2012 vs. 2000, -22.0 %). For Member States which joined the EU in 2004 and 2007 on the other hand, the agricultural labour input increased by 0.7 % compared with 2011, (2012 vs. 2005, +25 %; 2012 vs. 2000, +40 %).

Concerning the two EFTA countries for which agricultural labour input data were available, a decrease in 2012 compared with 2011 was reported by Switzerland (-1.7 %) and Norway (-4.0 %). In Croatia, the labour input decreased in 2012 (-2.0 %) compared with the previous year (2012 vs. 2005, -13.3 %).

Data sources and availability

Concepts and definitions: The estimates of the economic accounts for agriculture (EAA) published in this issue of 'Statistics in focus' are provisional. The compilation is based on information available up to the end of January 2012, which is also uploaded on Eurostat databases.

The results have been compiled by the national authorities in the Member States of the European Union, Norway and Switzerland, in accordance with the methodology of the EAA (which is close to the methodology of the national accounts, ESA95, but incorporates a number of changes to take account of the special features of the agricultural economy).

What is agricultural income? The income indicators relate to the income generated by agricultural activities (as well as inseparable nonagricultural, secondary activities) over a given accounting period, even though in certain cases the corresponding revenues will not be received until a later date. It does not, therefore, constitute the income effectively received in the course of the accounting period itself. Moreover, they are not indicators of total income nor the disposable income of farming households; in addition to their purely agricultural income, households often receive income from other sources (non-agricultural activities, salaries, social benefits, income from property). In other words, agricultural income must not be regarded as the income of agricultural households. In the present context, emphasis is placed on the Indicator A of agricultural income per work unit. Its development is presented as indices based on the figures of the previous year for short-term development, and on 2005 for medium. The indicator is defined as follows:

- **Indicator A:** Index of the real income of factors in agriculture, per annual work unit

This indicator corresponds to the real (i.e. deflated) net value added at factor cost of agriculture, per total annual work unit. Net value added at factor cost is calculated by subtracting from the value of agricultural output at basic prices the value of intermediate consumption, the consumption of fixed capital, and adding the value of the (other) subsidies less taxes on production. The detailed data may be found at the Eurostat website.

Other indicators: Two other indicators for agricultural income are also calculated:

- **Indicator B:** Index of real net agricultural entrepreneurial income, per unpaid annual work unit, and Indicator C: Net entrepreneurial income of agriculture.

Changes in recording of subsidies: Implementation of the Single Farm Payment (SFP) in the framework of the recent CAP started in several Member States in 2005. Since the Member States do not apply the SFP system in the same way, only limited comparisons of output values at basic prices, values of subsidies and value added amongst Member States and time periods (particularly 2004-2007) can be made. In general, implementation of the SFP resulted in a drop in output values at basic prices, and an increase in other subsidies on production, while values at producer prices and the income indicators A, B and C were not affected by this change.

Technical notes

- The data measured in real terms are obtained by deflating the corresponding nominal data with the implicit price index of gross domestic product (GDP) at market prices.

- The aggregates measured in real terms for the European Union as a whole are obtained by, first, deflating the nominal values at current prices recorded in the various Member States (by applying the implicit price index of GDP of the country concerned) and then by converting them into euro (at 2005 exchange rates for the medium-term analysis and at 2011 rates for the short-term analysis). The results are then added to obtain the real values for the European Union. It is on the basis of these aggregates that the developments for the European Union are calculated in real terms, which means that an 'EU deflator' is never explicitly used.
- In order to take account of part-time and seasonal work, agricultural labour or changes therein are measured in annual work units (AWU). One AWU corresponds to the input, measured in working time, of one person engaged in agricultural activities in an agricultural unit on a full-time basis over an entire year.

CONTEXT

The income indicators relate to the income generated by agricultural activities (as well as inseparable nonagricultural, secondary activities) over a given accounting period, even though in certain cases the corresponding revenues will not be received until a later date. It does not, therefore, constitute the income effectively received in the course of the accounting period itself. Moreover, they are not indicators of total income nor the disposable income of farming households; in addition to their purely agricultural income, households often receive income from other sources (non-agricultural activities, salaries, social benefits, income from property). In other words, agricultural income must not be regarded as the income of agricultural households.

In the present context, emphasis is placed on the Indicator A (see definition in Data sources and availability of agricultural income per work unit. Its development is presented as indices based on the figures of the previous year for short-term development, and on 2005 for medium.

From farm to fork - food chain statistics

The food chain usually starts on a farm, within the agricultural sector. While some goods are already processed on the farm, most food is processed or transformed subsequently within the manufacturing sector, before being distributed through wholesaling and transport activities. Consumers purchase the majority of the food they eat and the beverages they drink from a range of retail outlets (supermarkets, specialist food retailers, markets and stalls), but food and drinks can also be purchased from food service providers (restaurants, take-away outlets, cafés or bars).

Main statistical findings

From farm to fork - a statistical journey along the EU's food chain. In 2008 just over 48 million persons were employed within the EU-27's food chain; they worked in close to 17 million different holdings/enterprises - the majority of which (81.8 %) were agricultural holdings (Figure 1), often small in size. Together, all of the holdings/enterprises within the EU-27's food chain generated EUR 751 008 million of added value; the figures for value added equate to output minus intermediate consumption. The distribution of value added was quite different to that for enterprises/holdings, insofar as the relative importance of non-agricultural activities was considerably higher. For example, although food and beverages manufacturing enterprises accounted for just 1.6 % of the total number of holdings/enterprises in the EU-27's food chain in 2008, their contribution to the value added was much higher, at 26.0 %.

Structural differences across Member States

Particularly high numbers of agricultural holdings in eastern Europe. There was a considerable disparity between Member States as regards their respective structures within the food chain from agriculture through to retailing and food services.

The highest number of agricultural holdings was found in the eastern Member States - in particular, Romania and Poland - which were characterised by land fragmentation and subsistence holdings (rather than commercial farms). In contrast, the number of food and beverage manufacturing, wholesaling, retailing and service providing enterprises and local units peaked in the southern Member States - in particular, in Italy and Spain. For example, in Spain there was, on average, one local unit (single restaurant, bar or café) providing food and beverage services for each 157 inhabitants.

Agricultural and food and beverage manufacturing, wholesaling, retailing and service providing enterprises tended, on average, to be much larger, with higher market concentration in Germany and the United Kingdom, as well as some northern countries - such as Finland and the Baltic Member States.

Workforce size

Largest food and beverage retail workforce and food services workforce in the United Kingdom

The subsistence nature of farming in easter Europe is apparent, as more than 5 million persons were employed in the agricultural labour force in Poland and almost 6.5 million in Romania in 2008; these two countries accounted for 43.2 % of the EU-27's workforce of 26.7 million persons.

Germany had the largest workforce for food and beverage manufacturing (845 400 persons, 17.9 % of the EU-27 total). Spain had the largest number of persons employed in food and beverage wholesaling (350 600, 17.5 %) and specialised food and beverage retailing (254 700, 18.0 %); the latter accounted for more than one third (37.9 %) of those employed in Spanish food and beverage retailing (no other Member State reported a share above 30 %). There were 1.28 million persons employed in the United Kingdom within food and beverage retailing (17.3 % of the EU-27 total); this likely reflected a high propensity for supermarkets to engage part-time staff. A similar pattern was observed for food and beverage services, as the United Kingdom recorded the largest workforce in the EU (1.58 million, 21.6 %).

Agricultural holdings

'EU-27 food chain:'almost 14 million agricultural holdings Primary agricultural output includes, among others, products such as crops, animals ready for slaughter, or milk. These move down the food chain as agricultural output is processed/transformed into products ready for consumption. Processing can be relatively simple, such as grading or preserving, tinning and freezing, or may involve more elaborate transformations, such as the production of ready-to-eat meals.

The EU's common agricultural policy (CAP) was initially set-up in order to encourage an expansion in the volume of food being produced. The focus of the CAP has changed and there have been a number of policy reforms to redress problems associated with the over-supply of agricultural products, while considering the implications of intensive farming practices on the environment and food safety. The new CAP approach provides incentives for farmers to produce food in hygienic conditions, maintaining high standards of animal welfare, using environmentally-friendly production methods, while promoting a sustainable rural economy.

There were almost 14 million agricultural holdings across the EU-27 in 2007, with a regular farm labour force of more than 26 million persons.

Just over half of the EU-27's agricultural holdings in 2007 were specialised in producing crops while livestock farming was the next largest category (almost one third of all holdings). Crops also accounted for just over half of the EU-27's farm labour force.

Primary products of the food chain are split between crop products (such as cereals and vegetables), animals (cattle, pigs) and animal products (milk) - the relative importance of these was quite evenly distributed when measured in terms of gross output.

Food prices respond to changes in stocks that are essentially driven by the weather and its effect on harvests - see the peaks and troughs in wheat production. The quantity of wheat produced in the EU-27 peaked at 151 million tonnes in 2008 - before falling back to 138 million tonnes in 2009, while the production of potatoes fell from a high of 83 million tonnes in 2000 to a low of 57 million tonnes by 2006, before recovering somewhat to 63 million tonnes by 2009.

The production of specific agricultural and food products depends to a large degree upon climatic/geological conditions, the availability of land and water resources, and the level of imports; these factors help determine which type of farming is practised in each Member State.

Animal production is generally measured in terms of slaughtered carcass weight. Using this measure, pig meat was by far the largest category of meat production in the EU-27, at 22.0 million tonnes in 2010. This was 2.8 times as high as the production of bovine meat (from cattle) and more than 30 times higher than the production of sheep meat.

Almost one quarter (23.6 %) of the EU-27's production of cereals in 2009 was attributed to France, while Germany (16.8 %) and Poland (10.1 %) were the only other Member States to record double-digit shares.

Italy and Spain were the leading producers of fresh vegetables in the EU, each accounting for around one fifth of total production; the Netherlands was also specialised in producing fresh vegetables.

The United Kingdom and Germany recorded the highest production of drinking milk, and Germany's position as a leading producer ofdairy products was consolidated as it also recorded, by far, the highest production of cream.

As regards meat production, France and Germany were the largest producers of bovine meat in 2010 (19.3 % and 15.0 % of the EU-27 total), Germany and Spain were the biggest producers of pig meat (24.7 % and 15.4 %), while poultry production was widespread across most of the Member States. In contrast, sheep production was concentrated within the United Kingdom (39.2 % of the EU-27 total) and Spain (17.4 %), while Greece and Ireland were also relatively specialised.

BAKERIES DOMINATE MANUFACTURING

Bakeries account for over half of the food and beverage manufacturing enterprises in the EU.

Once crops/livestock or other products have been harvested/reared, they are generally processed, preserved or slaughtered in downstream manufacturing activities.

The manufacture of bakery and farinaceous products accounted for more than half of the food and beverage manufacturing enterprises in the EU-27 in 2008 (157 100); meat processing (which includes the slaughtering of animals) accounted for the next highest number (41 200); there were relatively few enterprises (3 800) engaged in the processing of fish and crustaceans in the EU-27.

There would appear to be a considerable difference in the average size of food and beverage manufacturing enterprises, as the relative importance of the subsectors was quite different in terms of their respective shares of value added (Figure 4). Relatively high shares were recorded for the manufacture of dairy products (9.4 %), meat processing (15.3 %) and beverages (17.4 %).

Imports

The EU is self-sufficient in meat, dairy products, cereals and beverages. Aside from its own primary production, the EU imports a range of agricultural products from non-member countries. Indeed, it is the world's largest importer of food - and a high proportion of imports come from developing countries. Farmers and food producers in non-member countries who wish to export their goods to the EU need to respect the food safety principles that apply for EU farmers and producers; checks are made on imports at European borders as food enters the EU.

Several EU Member States are self-sufficient in a range of food products (Table 6) - for example, meat or cereals; self-sufficiency indicates up to what point domestic production covers the needs or domestic use of each Member State. However, the EU imports a range of products to meet consumer demand - among which vegetables and fruit accounted for 26.5 % of total food and beverage imports in 2010, fish, crustaceans and molluscs (21.9 %) and coffee, tea and cocoa (17.4 %).

EU-27 imports of food and beverages were valued at EUR 78 254 million in 2010, with a trade deficit of EUR 5 095 million. The origin of EU-27 imports of food and beverages was highly diversified (as the top seven trading partners accounted for 43.9 % of total imports). The EU-27's two main partners for food and beverage imports in 2010 were both from South America: Brazil and Argentina.

Food transport within the EU

The majority of the food, beverage and tobacco goods that are transported within the EU travel less than 150 kilometres. The EU places great importance on the quality of food distributed to consumers. No matter the origin (EU farms or imports from further afield), the same rules are applied. Wholesaling and transporting play a vital role in the food chain, providing logistical services to move food and beverages between producers, processors and retailers.

One of the largest wholesaling activities concerns fruit and vegetables, employing 418 000 persons across the EU-27 in 2008 and generating about one sixth of the total value added within food, beverages and tobacco wholesaling.

Agricultural, forestry and fishing products, together with food, beverages and tobacco products accounted for just over a quarter of all the goods transported by road within the EU-27 in 2009 and for 18.1 % of the goods transported by road nationally; the majority of these goods were transported less than 150 km.

Changing consumption patterns

In recent decades, European consumers have increasingly turned to eating out. At the end of the food chain, consumers may choose to purchase food and beverages from specialist retailers (butchers, bakers), non-specialised outlets (supermarkets), or market stalls.

There were 7.4 million persons working within the food, beverages and tobacco retailing sector in the EU-27 in 2008. Non-specialised food retailing accounted for almost 80 % of the workforce and a higher share (86.2 %) of turnover.

In contrast, there were more specialised food, beverage and tobacco retail enterprises than there were non-specialised retailers. Among these, the most important activities (in terms of added value) included the sale of meat and meat products, and the sale of bread, cakes and confectionery.

Europeans have increasingly chosen to eat out or buy take-away foods during the last few decades. There were almost 1.5 million enterprises providing food and beverage consumer service activities in the EU-27 in 2008. Just over half were restaurants and 43.2 % were cafés or bars/pubs. The relative importance of restaurants was even greater in terms of their share of value added (58.1 % of the total).

PRICES AND EXPENDITURE

Restaurant and café prices rose faster than inflation during the last decade. Food prices may be affected by population growth and rising living standards (in emerging economies), which have changed global consumption patterns (meat replacing vegetables), while leading to increased demand for a range of food commodities. Within developed economies, urbanisation and higher living standards have also resulted in changes to consumption patterns, with more processed foods being eaten and people making more frequent trips to catering outlets. Competing uses for food output also influence prices - for example, farm output of crops (corn, sugar cane, rapeseed) may also be used as animal feed or for the production of biofuels. When food for human consumption is substituted by other uses then food prices are likely to rise.

Although the use of restaurants and take-away facilities has increased historically, this development stopped abruptly as the financial and economic crisis resulted in the volume of EU-27 household expenditure on catering services being reduced by 5.4 % between 2008 and 2009 - compared with a 2.4 % reduction for total food and beverages expenditure. Food and beverages accounted for 21.5 % of household expenditure in 2009, ranging from 17.4 % in the Netherlands to 34.1 % in Romania.

EU-27 consumer prices rose, on average, by 2.4 % per year between 2000 and 2010. Food prices increased at a slightly faster pace (2.9 % per year), with particularly large increases in 2001 and 2008; higher price rises were recorded for restaurants and cafés (3.2 % per year).

Data sources and availability

The statistical domain of the food chain is structured in 4 main areas:

- food consumption;
- from production to distribution: organic production, products with distinctive marks, production and sales of foodstuffs, prices and ancillary activities relating to packaging, transport and R & D in relation to food safety projects;
- inputs to the food chain: intermediate inputs to agriculture (such as feedingstuff, seed or fertilisers), primary production, extra-EU food imports (including information by partner);
- actors involved in the food chain: the number of local units and enterprises within manufacturing, distribution, transport and service activities, and the number of agricultural holdings.

Data sources

1. Survey on the structure of agricultural holdings (FSS)

Farm structure survey (FSS) data are used to collect information on agricultural holdings in the Member States at different geographic levels (Member States, regions, districts) and over periods, they provide a base for decision making in the common agricultural policy.

Two kinds of farm structure surveys are carried out:

- a basic survey (agricultural census) every 10 years,
- sample-based intermediate surveys.

2. Economic accounts for agriculture

This source provides information on income and expenditure in the agricultural sector. The data set includes statistics for the value of output (at producer and basic prices), intermediate consumption, subsidies and taxes, consumption of fixed capital, rent and interests, and capital formation. These accounts are a satellite account to the national accounts.

3. Structural business statistics (SBS)

SBS data have been collected within the legal framework provided by Council Regulation 251/2009. The SBS data set relates to annual enterprise statistics and provides information on a range of business-related issues, such as: business demography (births and deaths of enterprises), output-related indicators, such as turnover or value added; input-related indicators, such as labour input, purchases of goods and services, or investment. The data are presented using the NACE Rev. 2 classification. Information may be broken down according to enterprise size class and (at a more aggregated level of detail) according to different regions (as defined by NUTS).

4. Road and rail freight transport statistics

The legal basis for the collection of road freight transport statistics is Council Regulation 1172/1998. The figures are aggregated on the basis of sample surveys. The data are presented using the following units: tonnes, tonne-kilometres, vehicle-kilometres and numbers of journeys.

The legal basis for the collection of rail freight transport statistics is Council Regulation 91/2003.

Goods (for both road and rail) are classified according to the standard goods classification for transport statistics 2007 (NST 2007). The data cover road freight transport by heavy goods vehicles and exclude operations by small goods vehicles.

5. External trade statistics

External trade statistics are an important data source for EU decision-makers, being used extensively for multilateral and bilateral negotiations within the framework of the common commercial policy.

Extra-EU imports relate to goods entering the EU-27 from non-member countries; these statistics are collected directly from traders on the basis of their customs declarations. Extra-EU trade statistics do not record exchanges involving goods in transit, placed in a customs warehouse or given temporary admission.

The data are presented using the Standard international trade classification of the United Nations; currently, the fourth revision of theSITC is applied.

6. Annual national accounts

Household final consumption expenditure may be broken down according to consumption purpose (using the COICOP classification). The final consumption expenditure of households consists of expenditure incurred by residential institutional units on goods or services that are used for the direct satisfaction of the individual needs or wants or the collective needs of members of the community.

7. Price statistics: harmonised indices of consumer prices (HICP)

Harmonised indices of consumer prices (HICPs) are designed for international comparisons of consumer price inflation; they reflect price changes over time and provide the official measure of inflation within the EU and euro area. The coverage of HICPs is defined in terms of 'household final monetary consumption expenditure', by reference to the national accounts concepts of ESA95. Expenditures are classified according to the COICOP/HICP (classification of individual consumption by purpose adapted to the needs of HICPs).

DEFINITIONS OF VARIABLES AND INDICATORS

Number of enterprises: a count of the number of enterprises active during at least a part of the reference period.

Number of persons employed: the total number of persons who work in the observation unit (inclusive of working proprietors, partners working regularly in the unit and unpaid family workers), as well as persons who work outside the unit who belong to it and are paid by it (for example, sales representatives, delivery personnel, repair and maintenance teams); data relate to head counts. Note that within agricultural statistics the information in relation to persons employed may be provided either as a head count or alternatively as annual work units.

Context

The information presented is drawn from a range of different Eurostat databases. Many of these figures are collected together under the heading of 'Food: from farm to fork statistics', which brings together information on food products and the food sector which are considered relevant for food safety purposes. Primary reference data sources are also detailed in 'Data sources and availability'.

Large farm statistics

The utilised agricultural area (UAA) refers to the land used for farming; it includes arable land, permanent grassland, permanent crops and also kitchen gardens. The distribution of the UAA between large and small farms varies among countries but in most otf them a small number of farms occupy a high percentage of the agricultural area.

Utilised agricultural area

Less than 1 % of European farms occupy 20 % of the utilised agricultural area

In Malta, the 20 % of UAA covered by the largest farms is occupied by farms larger than 5 ha. In Slovakia, farms larger than 2 782 ha cover 20 % of the total UAA. These larger farms in Malta have an average UAA of 7 ha and in Slovakia of 3 934 ha.

In 20 countries, the larger farms are all in the biggest UAA class (100 ha or more). Only in Malta and Slovenia does the 80 % threshold lie below 30 ha.

In Bulgaria, the Czech Republic, Hungary, Estonia and Slovakia, the larger farms occupying 20 % of the UAA are all above 1000 ha. This pattern occurs in several of the new Member States; here the structure of the agricultural holdings is related to the particular ownership structure made up of large-scale corporate farms inherited from former state-owned cooperatives.

In Switzerland the average size of the larger farms (54 ha) is three times the average size of all farms (17 ha), while in countries such as Bulgaria (3 128 ha vs 6 ha), Romania (1 802 ha vs 3 ha) and Hungary (3 164 ha vs 7 ha), the average area of the larger farms is close to 500 times the average size of all farms.

In Slovakia, the Czech Republic, Hungary and Bulgaria the group of larger farms covering 20 % of the total UAA have an average area of over 3 000 ha. Within the old Member States (EU-15), Greece, with an average area of 64 ha for the larger farms, has the lowest average area while the United Kingdom is the only country showing an average area over 2 000 ha for the group of larger farms. This can be explained for the United Kingdom by the fact that larger farms specialise in grazing livestock extensively.

The group of larger farms, occupying 20 % of the total UAA, sometimes represents an extremely small proportion of all farms in a given country.

In Bulgaria, Hungary and Romania, only 0.04 % of the total number of holdings cover 20 % of the total UAA. Also in the Czech Republic, Germany, Estonia, Spain, Italy, Cyprus, Latvia, Lithuania, Poland, Portugal, Slovakia and the United Kingdom less than 1 % of the holdings cultivate 20 % of the total UAA.

The share of larger farms is only over 1 % in 14 countries, and only over 5 % in Norway, Luxembourg and Switzerland. The highest percentage is found in Switzerland; here 6.33 % of the farms cover 20 % of the country's UAA.

Adding up the number of larger farms (covering 20 % of the UAA observed for each country for the EU-27gives a total of 86 125 farms, which represents 0.6 % of all the farms in the 2007 FSS.

The larger farms do not have such a high average UAA per farm in the countries where they represent a greater percentage. On the other hand, with the exception of

Italy, Cyprus and Poland, in countries where the number of larger farms is less than 1 % of all farms, the average UAA per large farm is over 500 ha.

Standard gross margin

The FSS uses the standard gross margin (SGM) to measure the economic size of agricultural holdings. The SGM is the difference between the value of the agricultural output (crops or livestock) and the cost of inputs required to produce that output. The sum of all the margins per hectare of crop and per head of livestock in a farm is a measure of its overall economic size.

Overall, the total SGM of the agricultural holdings of the EU-27 in 2007 was 185 billion euro.

The average SGM per farm of the smaller farms that cover 80 % of the UAA varies from 1 057 in Romania to 123 942 in the Netherlands.

In 16 countries, the average SGM per farm of the smaller farms is under 20 000. All the new Member Statesexcept the Czech Republic belong to this group.

In the group of the larger holdings covering 20 % of the UAA, the average SGM per farm is highest in the Czech Republic (1.97 million euro) and the lowest averages per farm, all under 100 000, are observed in Malta, Greece, Austria, Slovenia and Ireland.

The difference in the average of SGM between the groups of smaller and larger farms is highest in Hungary, Bulgaria, Romania, Slovakia and Latvia. In all these countries, the average SGM per farm of the smaller holdings is less than 1 % of the average SGM per farm of the larger holdings. On average, the SGM of a large holding in Hungary is 448 times the average of a smaller Hungarian farm.

In countries such as Austria, the Netherlands, Belgium, Luxembourg and Norway the smaller holdings have an average SGM per farm that is at least 25 % of the average of the larger farms.

The SGM per hectare within the two groups of farms shows a different pattern from the average SGM. In 21 countries the smaller farms have a higher SGM/ha than the larger farms. In nine of those countries (Bulgaria, Greece, Spain, Italy, the Netherlands, Austria, Portugal, Romania and the United Kingdom) the SGM/ha of the smaller farms is more than twice the SGM/ha of the larger farms. In the Czech Republic, Estonia, Ireland, Latvia, Lithuania, Slovakia and Sweden, the larger farms' SGM/ha is slightly higher than the SGM/ha of the smaller farms.

The differences in the weight of the larger farms' SGM within the total SGM vary from 3 % in Austria to around 23 % in Latvia, Lithuania and Estonia.

In Slovakia, the Czech Republic and Ireland the larger farms, covering 20 % of the UAA, also make up 20 % of the SGM. In Sweden, Estonia, Latvia and Lithuania the share observed in each country exceeds 20 % of the SGM. In all the other countries, a lower share can be seen.

The total SGM of all the larger European farms covering 20 % of the UAA in each country makes up 11 % of the total EU 27 SGM but represents less than 1 % of the total number of farms.

Labour force

In 2007, there was a total of 11.7 million annual work units (AWU), the equivalent of 11.7 million people, working full-time in agricultural activities in the EU-27. In 2007, the larger farms accounted for 5 % of this total labour force.

In the smaller farms the difference in average AWU per holding between the European countries is quite limited, ranging from 0.3 AWU in Malta to 2.9 AWU in the Czech Republic. In the group of the larger farms there is a stronger contrast within the EU 27. The larger farms in Malta on average employ 1.8 AWU while in Slovakia the average per farm reaches 125 AWU.

Labour input of the larger farms as a percentage of the total labour force of all agricultural holdings is the lowest in Romania, where only 0.9 % of the labour force is occupied on 20 % of the UAA.

The 20 % of the UAA covered by the larger farms always represents less than 17 % of the total AWU. This percentage exceeds 10 % in only seven countries (Luxembourg, Switzerland, Denmark, Malta, Slovakia, Estonia and the Czech Republic).

The number of hectares of UAA per AWU is always greater in the larger farms than in the smaller ones. In the group of smaller farms the numbers are not very different, while in the larger farms they can vary considerably. The United Kingdom stands out with a figure of 594 ha/AWU, again related to the extensive rough pastures that make up the largest holdings. In Spain, Austria, Portugal, Romania and Sweden, for each full-time worker on the larger farms there is over 100 ha of UAA.

Livestock density

In the Farm structure survey the livestock unit (LSU) is used for measuring the quantity of livestock. The percentage of LSU in larger farms is 20 % or more of the total LSU in Finland, Slovakia and Estonia. On the other hand, the larger farms in Austria occupy only 1 % of the total LSU. Similar to the situation in the United Kingdom, the larger farms in Austria are made up of rough grazing with a very low livestock density. This is also the case in Bulgaria and Romania, where despite the higher average of LSU in the larger farms, their livestock per hectare (LSU/ha) is very low.

The larger farms of all EU 27 Member States have 10 % of the total LSU in the European Union.

The number of LSU per hectare is greater in smaller farms in most of the countries, with the exception of Estonia, Slovakia and Finland.

In Belgium, Malta and the Netherlands there is over 3 000 LSU per 1000 hectare in the small farms, the value in Malta being the highest at 5 157 LSU/1000 ha.

The larger farms are predominantly less intensive regarding livestock density. In Bulgaria, Austria and Romania the LSU/1000 ha is under 100.

Data sources and availability

The FSS 2007 was conducted in all 27 Member States, Norway and Switzerland. In Belgium, Luxembourg, the Netherlands, Finland, Sweden, the United

Kingdom, Norway and Switzerland, the FSS was conducted as a census; in all other Member States it was conducted as a sample survey. Regulation 571/88 states that the surveys must cover agricultural holdings where the UAA is at least one hectare or, if under one hectare, holdings which produce a certain proportion for sale or whose production unit exceeds certain physical thresholds. The same Regulation also sets, regardless of the threshold, the coverage to 99 % of the SGM.

Although the definition of an agricultural holding* is common to all countries, the thresholds that are applied by each country differ across Member States. This leads in the FSS to different coverage of small farms between countries. As these small farms contribute little to the total UAA in each country, the effect of this variance will be significant when comparing the number of farms, but not when comparing the UAA between Member States.

The size of an agricultural holding can be measured using several variables, such as the UAA, the economic output, the amount of livestock or even the labour force involved in the farm work. The major advantage of using UAA as a measure for the size of holdings is that the UAA is a very simple concept (that can be easily understood) and stable (no major increases or decreases in the amount of UAA).

One AWU corresponds to the work performed by a person engaged in full-time agricultural work on the holding over a 12-month period. The annual working time of such a worker is 1 800 hours (225 working days of 8 hours per day), unless there are different specific national provisions governing contracts of employment.

LSU: For each of the 23 categories of livestock surveyed in the FSS, a specific coefficient is established initially on the basis of the nutritional or feed requirement of each type of animal which converts the number of heads to an LSU number (1 LSU corresponds to 1 dairy cow or 10 sheep).

Context

The ongoing revision of the common agricultural policy brings into focus the size of farms in the European Union. To measure the size of farms, different variables can be taken into account such as the economic performance, the labour force working on the farm, the production or the area.

In this statistical article the utilised agricultural area (UAA) is chosen for classifying farms by size in each country (see data sources and availability for more information). This classification provides the basis for a comparative study on the significance of the larger farms in the EU 27, Norway and Switzerland, using the 2007 data from the Farm Structure Survey (FSS).

The UAA refers to the land used for farming; it includes arable land, permanent grassland, permanent crops and also kitchen gardens.

The distribution of the UAA between large and small farms varies between countries, but in general a relatively small number of farms occupy a high percentage of the agricultural area.

For each country the total UAA of all farms is broken down into eight classes according to the size of the holding's UAA.

Only a small proportion of the total UAA in Malta, Greece and Norway is used by farms with 100 ha or more, while in Bulgaria, the Czech Republic and Slovakia the great share of the total UAA is occupied by farms in this size class.

In order to compare the weight of the larger agricultural holdings in each country, farms were sorted cumulatively by their UAA size and then divided into two distinct groups: o smaller farms, whose UAA added up to 80 % of the UAA (under the 80 % threshold); o larger farms, making up the rest (20 %) of the UAA (over the 80 % threshold).

With this approach to analysis of the larger agricultural holdings, the definition of 'large farm' depends not on a uniform threshold, but reflects the different distribution in the different countries. This means that the measurement of the size of the farms is always relative to the distribution of each country's UAA among the holdings.

6

Budgeting

A budget is a list of all planned expenses and revenues. The origin of this word came from French word bougette. It is a plan for saving, borrowing and spending. A budget is an important concept in microeconomics, which uses a budget line to illustrate the trade-offs between two or more goods. In other terms, a budget is an organizational plan stated in monetary terms.

In a nutshell, the purpose of budgeting is to:

1. Provide a forecast of revenues and expenditures, that is, construct a model of how our business might perform financially if certain strategies, events and plans are carried out.
2. Enable the actual financial operation of the business to be measured against the forecast.

BUSINESS START-UP BUDGET

The process of calculating the costs of starting a small business begins with a list of all necessary purchases including tangible assets (for example, equipment, inventory) and services (for example, remodeling, insurance), working capital, sources and collateral. The budget should contain a narrative explaining how you decided on the amount of this reserve and a description of the expected financial results of business activities. The assets should be valued with each and every cost. All other expenses are like labour factory overhead all freshmen expenses are also included into business budgeting.

CORPORATE BUDGET

The budget of a company is often compiled annually, but may not be. A finished budget, usually requiring considerable effort, is a plan for the short-term future, typically one year. While traditionally the finance department compiles the company's budget, modern software allows hundreds or even thousands of people in various departments like operations, human resources, IT, etc. to list their expected revenues and expenses in the final budget.

If the actual figures delivered through the budget period come close to the budget, this suggests that the managers understand their business and have been successfully driving it in the intended direction. On the other hand, if the figures diverge wildly from the budget, this sends an 'out of control' signal, and the share price could suffer as a result.

Effect of Climate Change and Drought

Although the country produces its own maize, because of climate change and the resultant drought in some parts of the country, most farmers are only harvesting enough crop for domestic consumption.

"One packet of maize flour has shot to almost two dollars when we always bought it for slightly above a dollar. Without maize flour, the country is facing a severe food crisis that has not been experienced since the 90's," explains Tim Njiru, a maize trader in Eldoret county in the Rift Valley.

But with a recent budget allocation of almost $112 million for agriculture, maize shortages may soon become less frequent in years to come. For the first time in the country's history, the agricultural sector has received a budget allocation of almost $112 million. The allocation to agriculture leaped from a paltry four per cent to nine per cent.

At nine per cent, the budget allocation is only one percentage point shy of meeting the Comprehensive Africa Agriculture Development Programme (CAADP) policy framework. CAADP requires that countries signatory to the agreement allocate at least 10 per cent of the national budget to agriculture.

Until now the sector has been underfunded despite its significance to sustainable human development, a situation that has further been complicated by extreme and unpredictable climatic conditions.

Kenya's economy is predominantly dependent on agriculture, according to the ministry of agriculture, the sector directly contributes an estimated 26 per cent of the Gross Domestic Product (GDP) and an additional 25 per cent indirectly.

According to Titus Warimi, an agricultural officer in the Rift Valley region, which is the country's breadbasket, "insufficient funds hamper research into various agricultural products. With the persistent and drastic climatic changes, it is imperative to venture into crop options that can flourish under the circumstances."

According to Kenya Food Security Meeting (KFSM) the country's main coordinating body that brings together various stakeholders to ensure that the country is food secure, the changing weather patterns will continue to impact heavily on the country's ability to feed its people.

"It will be difficult for the country to make any money from the agricultural sectors if farmers are hardly making any money," adds Warimi.

Ruth Ngige, a small-scale farmer in central Kenya echoes Warimi's sentiments. "Farmers are plowing back much of their profits into buying farm inputs, this is due to lack of subsidies in fertilisers and seeds. The products then become too expensive and customers therefore buy only what they really need to survive," she says.

It is because of the importance of agriculture on the continent that the African Union's (AU) NEPAD established CAADP in July 2003.

High growth

Under CAADP, the Common Market for Eastern and Southern Africa has been hard at work to ensure that member states move towards the attainment of Millennium Development Goal One (MDG1) to cut hunger and poverty by half by 2015 and to ensure environmental sustainability.

CAADP is the highest policy framework for the development of agriculture in Africa and its overall goal is to help African countries reach a higher path of economic growth through agriculture-led development which eliminates hunger, reduces poverty and food insecurity and enables expansion of exports.

"Eighteen member states have initiated the CAADP process by nominating CAADP focal points and 13 of these members, Kenya included, have launched the CAADP process," explains Prof Mary Abukutsa- Onyango, a lecturer in Jomo Kenyatta University of Agriculture and Technology, and a pioneer of extensive research into traditional vegetables like African eggplant, nightshades and cowpeas.

Planning and Budgeting for 2012 US Farm Bill

In the USA, discussions are slowly grinding forward on what the 2012 Farm Bill might look like. However, budgetary circumstances are a significant dark cloud hanging over the debate.

If you thought Common Agricultral Policy (CAP) is difficult to get your head around, try the Farm Bill. The Food, Conservation, and Energy Act of 2008 is the most recent Farm Bill covering farm and food policy. The Farm Bill is made up of 15 titles. Some 97% of the Farm Bill's $57bn annual spend is based on four titles: nutrition (67%), farm commodity support (15%), conservation (9%) and crop insurance (8%).

In order to get support from urban politicians, the major component of the farm bill is the nutrition programme, including food stamps. A whopping one in five Americans participate in at least one of the United States Department of Agriculture's (USDA) nutrition programmes. Everyone is in agreement that the single most important factor in developing the next Farm Bill is the budget. The other factors include policy, politics, and trade.

Unlike Europe, the Farm Bill in the USA is written by the politicians on the Senate and House of Representatives Agriculture committees. They are, therefore, critical in the discussions, none more so than the respective chairs, Senator Debbie Stabenow and Representative Frank Lucas.

While agriculture benefits from being largely bipartisan, discussions on the 2012 Farm Bill have been deliberately slowed a bit to allow the fiscal 'freshmen' get an understanding of how things work. Last year's mid-term elections saw the Republicans take the majority in the House of Representatives and make serious inroads into the Democrats' majority in the Senate.

The new Republicans on the Hill have a strong 'Tea Party' mandate to cut the federal spend and, with further Senators facing elections next year, there is a distinct possibility that incumbent Democrats may vote with the Republicans on spending cuts before going in front of the electorate.

The US debt limit will be reached in August with the consensus that the President and Congress can come to an agreement which will see the Republicans agree to the debt ceiling being raised with the proviso of substantial spending cuts.

So, where will the axe fall? Let us start by putting the Farm Bill cost in perspective. The Farm Bill annual spend is $57bn. The overall federal spend is $3.5 trillion, the bulk of which is defence, health care and social security.

This year, agriculture has already seen a $4bn cut in crop insurance. In fact, agriculture groups have accepted the fiscal reality and shown real leadership in presenting a common message that cuts are needed but such cuts should be equitable and not target agriculture over other areas.

The preference is for the agriculture committees to decide where the cuts fall and not those on the budget committees.

Cuts in the order of $10bn over a decade for programmes with a budget line have been recommended by the Presidential Commission and by Senate budget chairman Kent Conrad. However, reflective of the Tea Party agenda, the House of Representatives recently passed a bill containing $48bn in cuts to farm programmes over ten years. That proposed cut has no chance of making it into legislation, but agri groups, and political leaders from the Senate Agriculture Committee have raised concern that mentioning such figures moves the middle ground in agriculture spending cuts closer to $15bn.

Direct payments

Direct payments are the number one target. With an annual spend of $5.2bn amounting to 74% of the Title 1 Commodity spend, it appears that agri groups would accept their elimination provided the safety net is maintained and the highly popular crop insurance programmes are strengthened.

There is also a suggestion among some commentators that provisions be built for a possible re-introduction should farm incomes fall significantly. In other words, the bean counters should take cognis-ance of the fact that commodity prices won't stay high forever.

Lest anyone get ideas, direct payments only make up less than 10% of farm income here and the government-subsidised crop insurance, allied to future trading, is the preferred model for risk management.

One of the critical factors in weighing up the future of particular programmes is whether the programme has a budget line.

If it doesn't, you would be getting worried. Now, without delving too far into the technical details, any spend within the Farm Bill will either have what is called a baseline or it won't. A baseline provides assurance about the long-term prospects of spending for a particular area. Without a baseline, in the current economic circumstances, the area in question is looking like it will be chopped.

Some 37 programmes in the current Farm Bill do not have a baseline beyond 2012. The titles with the most such programmes are energy (eight programmes), conservation (five), nutrition (five) and horticulture and organic agriculture (five). Within these, the Agriculture Disaster Assistance Program, the Wetlands Reserve Program and the Biomass Crop Assistance Program account for 75% of the $9bn to $10bn total.

In the past, touching the nutrition title was unlikely given its urban support and the sensitive nature of cutting from the neediest. However, in the current climate, everything is up for discussion and with 67% of the budget, a small cut in nutrition programmes would offset much of the needed reduction in Farm Bill spending.

ORGANIZING FOR SUCCESS

Agribusiness firms are continually faced with the moving, handling, processing, and storage of products from producer to consumer. The coordination, planning, and implementation of such activities are likewise integral to the ultimate success and survival of the firm. Activities such as these generally referred to as "logistics," represent a sizable portion of the total price for most agricultural products.

In a an article in the trade magazine *The Private Carrier*, Data Resources, Inc., estimated that in paper, furniture, food, chemicals, and lumber, logistics related costs represented over 20 percent of the delivered product price. Hence, by minimizing such costs there exists an opportunity to obtain a competitive advantage. The principal means of controlling logistics activities and insuring efficient, cost-effective operation requires understanding the important issues emerging in logistic operations.

To this end, this chapter first reviews logistic goals and performance criteria, and then discusses current issues affecting the achievement of such goals.

Logistic Goals

In the simplest terms, most businesses purchase inputs in some form or fashion, transform or alter them in some way, and then sell the transformed product as output. Thus, two principal product flows are apparent as inbound and outbound movements. While individual aspects concerning the timing, form, and manner of coordinating these product movements may differ for different types of businesses, there are fundamental standards that all businesses engaged in this type of activity consider.

For both types of product movements, dependable, accurate, error free, logistic service is essential to the successful coordination of any agribusiness operation.

Regardless of whether the logistics of inbound and outbound product movements are controlled within the company or contracted to an outside party, service and performance are critical. Service that is subject to frequent delays and scheduling problems is unacceptable in the current business environment. Likewise, products must be transported free of damage or injury on a consistent basis.

Providing this type of service, while controlling cost, is the primary goal for logistics and agribusiness managers. Several avenues for achieving these types of goals have recently been discussed in the literature. While no single strategy will fit the needs for all agribusinesses, certain issues dominate the discussion as to what would be included in a strategy.

Emerging

Safety

Safety is one of the emerging "musts" in logistics operations. With the growing public concern over transportation of hazardous materials and increased state and federal regulation requirements, safety programs must address these concerns. A recent article in The Private Carrier, Lend Lease, Inc., writes about the safety program implemented by one of their clients, Liquid Air Corporation.

The safety program developed for this company was designed to meet all government requirements for driver certification, testing and education, and also provided driver incentives for highway safety performance. In addition, the company updated the equipment maintenance facilities in order to keep safe, reliable equipment operating efficiently. All drivers are now required to pass the

Commercial Driver License exam, which tests both skill and knowledge. In addition, companies that transport hazardous materials are required to implement a Controlled Substance testing program for employees. Safety seminars dealing with driver training, handling instructions and regulation changes also update drivers on safety issues.

Emphasis on Quality

The emphasis on quality has historically been placed on the end product with only casual concern for the production process. However, today's agribusiness managers must understand the importance of integrating quality standards into all aspects of the business operation. This has resulted in the phrase "Total Quality Management" as a characterization of the logistics and general manager's goal.

The impetus for implementing a quality program resulted from the interrelated benefits associated with service, performance, and customer satisfaction. The critical question then is how to define "quality" and determine a list of priorities or goals.

Conceptually, the term "quality" can be applied to any measure of logistics service for performance.

The ultimate goal, however, is customer satisfaction. Given the subjective nature of the term quality, measurement of quality programs is sometimes difficult. However, some examples of commonly measured quality variables which are less subjective include: (a) on-time pickups and deliveries, (b) order cycle time, (c) error free transactions, (d) goods delivered free of damage, (e) equipment availability, and

(f) accurate billing. Many other possible variables measuring customer satisfaction could also be included, depending on the nature of the specific business.

The important point, however, is that quality measures should be specifically defined and prioritized for each business. Once determined, quality measures can then be monitored and evaluated to determine the success of implemented quality programs.

Several key considerations are often mentioned when devising a successful total quality management program. Richard D. Armstrong argues in a an article in *The Private Carrier* that the single most important requirement for successful implementation of quality programs is the support by top managers. If the top managers are enthusiastic and fully supportive of the quality program, then its chances of success are greatly increased.

It is also important to begin a quality program that is easily implemented with benefits quickly noticed, thereby reinforcing belief in the program. Eventually, however, defining and achieving a long-term commitment to quality is necessary.

Environmental

In addition to the business related activities involving the coordination and control of products from producers to consumers, logistics managers must also consider the intervening presence of the government. More important for the logistics and transportation industry is the ever-growing presence of government regulating and influencing the way it operates. Regulation by state and federal agencies influences logistics activities in several different forms, but the growing emphasis involves environmental concerns.

A recent study from Transport Topics estimated that environmental cost for the trucking industry will increase from $1 billion to $12 billion by the year 2000. This represents a sizable increase in costs associated with environmental compliance.

In addition, the understanding and interpretation of added legislation is becoming increasingly difficult as the range of regulated issues becomes broader. In addition to past state and federal legislation covering air pollution, water pollution, and transport of hazardous materials, new legislation is resulting in regulation involving oil and tire recycling, and more stringent requirements involving acceptable routes for transporting hazardous materials.

Environmental Program

One useful strategy is to develop an environmental program, much the same as a safety or quality program. The environmental program need not be as complex and can simply consist of a small staff or assigned personnel whose primary responsibilities are staying abreast of current regulations associated with all business activities and informing all employees. The necessity for environmental compliance is not derived from gains in efficiencies or profits, but rather a simple fact of doing business.

INFORMATION AND TECHNOLOGY

One of the current opportunities for realizing a competitive advantage is from the information and technological capabilities emerging in logistics operations. Information facilitates practically all areas of management, including implementation and measuring the performance of safety, quality, and environmental programs. The key to information accessibility is technology. Logistics operators today have a wide variety of technologies available for many different types of information and control enhancements.

One broad category of coming information technology is called Intelligent Vehicle Highway Systems (IVHS) that provides communication, vehicle identification, warning and control systems, prompt operations information, driver and dispatcher correspondence, route guidance, and monitoring systems. Given this type of information capability, the potential management uses are numerous. Communication between truck drivers and dispatchers helps avoid delays and scheduling problems, as well as providing upto-the-minute information on driving conditions, weather, traffic, highway construction information, and detours. Vehicle identification would facilitate easier compliance at weigh stations by identifying and inspecting individual trucks without stopping the vehicle. Performance information is also available to drivers and managers to help assess such measures as fuel efficiency and truck performance.

The degree to which technology is embodied and applied in daily business activities will depend largely on the company and the characteristics associated with it. Certainly, larger companies will be able to take advantage of the economies of size by spreading the cost of investing in new technology across more production units. Smaller companies may have trouble justifying the large initial cost of advanced technology and its application to logistics operations as essential for all successful logistics management.

The shape of the transportation and logistics industry is changing rapidly; hence, any agribusinesses involved with these activities changes as well. The result is increased demand for information and education concerning issues affecting these types of businesses. We have mentioned and discussed the more current and popular issues confronting logistics operations in an attempt to help managers prepare for the future.

Indian Society of Agribusiness Professionals (ISAP) is a non-government, non-profit organisation incorporated in 2001, under Section 25 of the Indian Companies Act. It is a network of agriculture and allied sector professionals in India and developing countries. It is a growing network encompassing over 15,000 registered associate members, comprising 1500 Agri-Experts, 525 Partner NGOs, over 824 individual users and more than 1050 researchers.

Management of OCP Project

Rains in the OCP project area of Karnataka have been unduly delayed and the sowings took off primarily in the second of July only. In the mean time, many farmers who had banked primarily on the variety BSMR 736, a late maturing variety wanted to switch over to TS3R, a relatively shorter duration variety following the advice rendered by the Project management in the form of pamphlets distributed as an alternate contingent plan. In order to further support this switch over, Mr Rajiv Dar, Chairman ISAP unilaterally decided that all project farmers would be provided seeds of TS 3R at a cost of only Rs 100/- for one acre, as against the actual market cost of Rs 408/- (fixed by UAS Raichur). This was welcome by farmers as evidenced by the fact that 500 farmers came forward to purchase this seed in Gulbarga district alone. He also announced that during current season, ISAP would provide service (on actual cost basis) towards plant protection of pigeonpea crop, to all farmers in the OCP-AES project geography, as and when the need for this is felt by any farmer. Developing indigenous plant based insect repellants as an aid to integrated pest management at farmers' household level is again a new feature of the training being imparted on reducing production costs.

A major events during the current month was a visit to the OCP-AES project area by a high-power team of OCP-INRA from Morocco, comprising Mr. Mohamed Belmahi, Advisor to Chairman - Groupe OCP and President - OCP Foundation, Prof. M. Badraoui, Director General, INRA, Morocco, Ms. Fatiha Charradi, President Director General, OCP Innovation Fund for Agriculture and Ms. Hassina Moukhariq, Food Security Program Manager, OCP Foundation Research Center.

During the later part of the visit the team was joined by Dr S A Patil, Chairman Agriculture Commission, Govt of Karnataka who is also an advisor to the OCP management. It was a landmark visit characterized by intensive, in-depth evaluation of the project activities and project farmers in all the three districts. Though the sowings had not taken place in many places due to delayed rains, happily, it began to rain during the visit of the dignitaries from Morocco.

Enthusiasm Among Farmers

A notable feature of the visit was the high level of enthusiasm shown by farmers not only in raising their own yields but to act as catalyzing agents for educating their fellow farmers in adjoining villages and help them achieve the kind of yield advances in pigeonpea production realized by them under this project. What was further remarkable is the consciousness exhibited by the project farmers about the health of the soil and the level of plant nutrition which is responsible for higher productivity from their farms. Most of the farmers visited said that they had realized double the yield of pigeonpea during 2010 as compared to ever before.

New cultivation practices of pigeonpea introduced by the OCP-AES project like transplanting and dibbling of pigeonpea had helped them a good deal in this endeavour. As a result, three nurseries have been set up with three farmer groups- one in each district under the project. The dialogues with farmers were so charged with success that Prof. Badraoui exhorted the farmers to attempt another doubling of productivity this year. The farmers said that if it rained well they would certainly like to try it.

The team was impressed by the formation of the "farmers' groups" and the impeccable record keeping of all the activities of individual farmers covered under the project. During the visit Prof. Badraoui inaugurated a "Low Cost Farm Machinery Center" started by a farmer group in Ankalga village of Afzalpuir block in Gulbarga district. While the major funding for this center has come from the project, the farmers themselves have shown ownership of the activity by contributing part of the cost from their own resources.

Again, the water harvesting in farm ponds and its possible recycling resources created with selected farmer groups impressed the visitors greatly. Other group activities by project "farmers' groups" initiated/to be initiated this year include seed production, small dal mill and modern nursery facilities.

The visitors were overwhelmed with the warm reception and hospitality on their welcomed by the project farmers. The Team expressed immense pleasure to watch a folk dance arranged by the predominantly tribal population in village Garda and they were also drawn into the dance by village folks. In response, the visitors showered gifts on some of the farmers and children in the villages.

The "water filters" gifted to farmer families would kindle in them an urge for family health just like the soil health program being carried out under the project. The kids of the village received reading and writing materials to encourage the urge for education.

The orgznal structure of firms affects how employees perform their work & make decisions in various situations. Firms usually organize according to:

(a) **Business Function:** Here firms tend to have large, separate dpts–e.g. sales, pdn, & purchasing dpts,

(b) **Product Lines:** Others organize along pdt lines – e.g. feed, seed, fertilizer, etc. Here all activities associated with the pdt are done by a separate mgmt team.

(c) **Geographic Location:** Organizing by geographic location allow all activities of a firm in northeastern Arkansas to be handled by the Jonesboro office.

Controlling: Measuring Organization's progress toward goals set in planning stage. Found at all levels and include measuring progress of work in entire dept, or orgzn. Progress levels is measured against standards set in planning and the standards must not be set too high/low

The situation determines type & amount of corrective action to be taken. In some situation only minor adjustments is taken, in others more drastic action is required. Rule is to properly identify the problem and seek ways to correct it.

Directing: This implements planning, organizing, and controlling to transform the plan into a reality – i.e. mager meshes the plan, organization, and controls with human, physical resources, to accomplish the objectives quickly & efficiently. It consumes about 90% of mager's time.

The success of directing process is rest on workers who do the work. If employees are involved in planning they are willing to work toward its success. Properly managed employees can make poor plan succeed

Choosing a Legal Structure

A business is a legal entity that takes on a set of characteristics and requirements separate from the individuals who form and operate the business. State law determines the legal requirements associated with each form of business. Federal and state law have their own sets of tax requirements, and business owners need to be aware of both sets of laws when choosing a business structure.

It is important to explore and understand the full range of options available to an entrepreneur. The choice of legal structure in India can either be a for-profit entity (i.e. a sole proprietorship or a company or a partnership firm or a limited liability partnership etc.) or not-for-profit entity (i.e. a society or a trust or a company registered under Section 25 of the Indian Companies Act, 1956 etc.) or even a hybrid structure (i.e. affiliate for-profit and not-for-profit entities functioning together as a cohesive business unit and interacting with each other on an arm's length basis whileleveraging their respective synergies).

Choice of Legal Structure

The choice of legal structure, whether it should be a stand-alone entity or a hybrid structure, or whether for-profit entity or not-for-profit entity, should reflect the business plan of the entrepreneur, and not necessarily vice versa. Entity structures are intended to foster and facilitate entrepreneurship and factors such as business operational model, regulatory and tax considerations, transaction costs, sector specificity of activities (e.g. microfinance, agriculture etc.), type of funding sought, dynamics between and amongst the various promoters and funders, proposed mechanism for profit sharing or distribution, governance considerations, limited liability aspects etc. collectively govern the choice of legal entity.

Sole Proprietorship

A sole proprietorship is the simplest of the available business structures. Individuals who wish to form a business as a sole proprietorship may apply to the Internal Revenue Service online to obtain an Employer Identification Number (EIN). This number is required in order to open a business checking account in the business' name.

Individuals who wish to reserve a business name so that it is not used by any other business in a given county may do so at the county Register of Deeds office. The individual will be instructed to search the Grantor/Grantee index over the previous five years to make certain that the name has not been previously reserved.

Those who wish to reserve a business name on a statewide basis may contact the North Carolina Secretary of State's office. On online database is available to determine whether or not a business name is already reserved. Individuals who operate as a sole proprietorship have the opportunity to file a Schedule C or Schedule F as part of their tax return. Any profit from the operation of the business is taxed at the individual's personal tax rate. A loss from the operation is used to offset the taxes due on other income earned..

Partnership Operations

An unincorporated partnership operates in much the same way as a sole proprietorship with regard to legal and tax liabilities. The partnership must apply to the Internal Revenue Service to obtain an EIN. The partnership will need to open a separate partnership bank account.

Members of the partnership draw up a partnership agreement that identifies each partner in the enterprise, the percentage ownership of each and any specific information that better defines each partner's role, such as a list of job responsibilities or investment participation.

The partnership completes the appropriate schedule, and the division of the profits (or loss) are reported as income (loss) on each partner's individual tax return.

As is the case with a sole proprietorship, the partnership should consult the appropriate legal and insurance professionals to determine the best way to avoid unnecessary legal liabilities resulting from the operation of the business.

Increase Investment in Farming Ventures

Minnesota and North Dakota during the past five to ten years has been rapidly developing elsewhere as well. Producers of various commodities have begun to increase investment in their farming ventures, not by the traditional method of expanding production, but rather through investment in initial or first stage processing of their agricultural commodities and through the second stage marketing of their commodities. To some extent, further marketing and processing have been driven by low commodity prices, but, in general, these trends reflect the philosophy of a growing number of producers that producer-owned markets, whether for processing or otherwise, are essential for farmers' continued existence in agricultural production.

Value-added Agribusiness

Many types of value-added agribusinesses have been formed to further process raw agricultural commodities. Examples of such value-added agribusinesses include processing corn into sweeteners and ethanol; processing corn or soybeans into feed for hog production, fish production, and chicken and egg production; processing soybeans into structural board products; extruding oil and other related products from soybeans; processing hogs and marketing meat products; processing cattle and marketing beef products; and a number of other ventures. The capitalization of these individual ventures has ranged from $500,000 to $200 million, with most projects costing from $3 million to $10 million.

Other businesses, which are primarily cooperatives, have formed to market raw agricultural commodities. These businesses usually focus on quality characteristics of a specific commodity or commodities for which there is a limited or nonexistent market.

For example, farmers have formed marketing cooperatives in which members produce specific varieties of high oil corn as well as corn and soybeans with specific genetic traits.

The marketing cooperatives enter contracts with end users for these identity-preserved commodities. Other commodities, such as specialty crops and fruits and vegetables, have also been marketed in this manner.

One of the most important reasons why these businesses are being formed is to return profits to the farmer producers who provide the investment. Secondary benefits include the association of a number of independent farmers into a common enterprise, as well as the creation of businesses with stable and long-term ties to the local communities.

In turn, profits from these businesses are returned to owners who most typically are members of the community as opposed to investment groups living far away from the community.

Anyone who has visited Renville, Minnesota, has noticed the bustling downtown business district as well as the newly constructed community center that serves as a meeting hall for various cooperatives and their members. The City of Renville has also started an industrial development park that is intended to attract agribusinesses, rather than trying to lure high-technology or medical device industries.

Currently, Golden Oval Eggs, an egg production cooperative and MinAqua, a tilapia fish production cooperative are located in the city's industrial park and together they employ almost 100 people. As a rural economic development tool, locally owned value-added agribusinesses should be one of the top attractions for rural communities.

BUSINESS PLAN FOR AGRIBUSINESS PROJECT

During the life of an agribusiness project, a business plan takes many different shapes and forms. In their most basic form, business plans may start on a napkin that reflects a discussion among local producers at a local café as to additional margins that can be received on a specific project, the estimated costs of the project, and the likely capital cost per producer. The napkin business plan captures the basic premises—project costs, profit potential, and the capital contribution required for investor or per unit commodity contributed to the project—on which a successful business can be built. The napkin business plan created at the coffee shop is often the basis from which a more formal business plan is developed.

Business Idea

The further development of a business plan typically involves verifying assumptions on which the business idea is based. The potential market for the products and the anticipated project costs are usually examined on a more detailed basis by using consultants who outline the project in a more detailed manner and verify specific costs.

Costs typically include land, buildings, and equipment costs; organizational costs; consultant and legal costs; and regulatory costs necessary to start and operate a project. In addition to the initial cost of establishing the venture, it is important to further quantify the profit potential by identifying operational costs of the venture such as labor, utilities, inputs, debt financing, and other ongoing operational costs.

Revenue assumptions about the project must also be quantified and verified. An assessment of market or sales price paid for the commodity must be made. Frequently, this investigation takes two steps: (1) identification of pricing that is commonly paid and (2) determination of whether the market for the pricing being used is actually available for the products of the agribusiness venture. For example, a business plan for a bean processing facility would need to identify the price being paid for the processed beans and determine the potential for the markets to purchase as much processed bean products as will be produced and sold by the venture.

Of all the choices you make when starting a business, one of the most important is the type of legal structure you select for your company. Not only will this decision have an impact on how much you pay in taxes, it will affect the amount of paperwork your business is required to do, the personal liability you face and your ability to raise money.

Mark Kalish is co-owner and vice president of EnviroTech Coating Systems Inc. in Eau Claire, Wisconsin, a company that applies powdered paint through an electrostatic process to items ranging from motorcycles to musical instruments. Kalish has also been involved with a number of other start-up businesses, both as an owner and in various management positions.

The answer to the question of "What structure makes the most sense?" depends, he says, on the individual circumstances of each business owner. "Each situation I've been involved with has been different," he says. "You can't just make an assumption that one form is better than another."

It's not a decision to be entered into lightly, either, or one that should be made without sound counsel from business experts. It's important for business owners to seek expert advice from business professionals when considering the pros and cons of various business entities.

That advice can come from a variety of sources, ranging from the no cost/low cost, such as the SBA or the Service Corps of Retired Executives (SCORE), to pricier attorneys and accountants who can serve as valuable sources of information throughout the life of your business.

Types of Business Entities

The type of business entity you choose will depend on three primary factors: liability, taxation and record-keeping. Here's a quick look at the differences between the most common forms of business entities:

A sole proprietorship is the most common form of business organization. It's easy to form and offers complete managerial control to the owner. However, the owner is also personally liable for all financial obligations of the business.

A partnership involves two or more people who agree to share in the profits or losses of a business. A primary advantage is that the partnership does not bear the tax burden of profits or the benefit of losses-profits or losses are "passed through" to partners to report on their individual income tax returns. A primary disadvantage is liability-each partner is personally liable for the financial obligations of the business.

A corporation is a legal entity that is created to conduct business. The corporation becomes an entity-separate from those who founded it-that handles the responsibilities of the organization. Like a person, the corporation can be taxed and can be held legally liable for its actions. The corporation can also make a profit. The key benefit of corporate status is the avoidance of personal liability.

The primary disadvantage is the cost to form a corporation and the extensive record-keeping that's required. While double taxation is sometimes mentioned as a drawback to incorporation, the S corporation (or Subchapter corporation, a popular variation of the regular C corporation) avoids this situation by allowing income or losses to be passed through on individual tax returns, similar to a partnership.

A hybrid form of partnership, the limited liability company (LLC) , is gaining in popularity because it allows owners to take advantage of the benefits of both the corporation and partnership forms of business. The advantages of this business format are that profits and losses can be passed through to owners without taxation of the business itself while owners are shielded from personal liability.

When making a decision about the type of business to form, there are several criteria you need to evaluate. Kalish and EnviroTech co-owner John Berthold focused on the following areas when they chose the business format for their company:

Legal liability. To what extent does the owner need to be insulated from legal liability? This was a consideration for EnviroTech, says Kalish. He and Berthold had a hefty investment in equipment, and the contracts they work on are substantial. They didn't want to take on personal liability for potential losses associated with the business. "You need to consider whether your business lends itself to potential liability and, if so, if you can personally afford the risk of that liability," Kalish says. "If you can't, a sole proprietorship or partnership may not be the best way to go."

Carol Baker is the owner of the company Corporation, a firm based in Wilmington, Delaware, that offers incorporation services. She points to the protection of personal assets as "the number-one reason our clients incorporate. In case of a lawsuit or judgment against your business, no one can seize your personal assets. It's the only rock-solid protection for personal assets that you can get in business."

Tax implications. Based on the individual situation and goals of the business owner, what are the opportunities to minimize taxation?

Baker points out that there are many more tax options available to corporations than to proprietorships or partnerships. As mentioned before, double taxation, a common disadvantage often associated with incorporation, can be avoided with S corporation status. An S corporation, according to Baker, is available to companies with less than 70 shareholder returns; business losses can help reduce personal tax liability, particularly in the early years of a company's existence.

Cost of formation and ongoing administration. Tax advantages, however, may not offer enough benefits to offset other costs of conducting business as a corporation.

Kalish refers to the high cost of record-keeping and paperwork, as well as the costs associated with incorporation, as one reason that business owners may decide to choose another option—such as a sole proprietorship or partnership. Taking care of administrative requirements often eats up the owner's time and therefore creates costs for the business.

It's the record-keeping requirements and the costs associated with them that led Kalish to identify the sole proprietorship as a very popular form of business entity. It's the type of entity in place at his other business, Nationwide Telemarketing.

"I would always take sole proprietorship as a first option," he says. "If you're the sole proprietor and you own 100 percent of the business, and you're not in a business where a good umbrella insurance policy couldn't take care of potential liability problems, I would recommend a sole proprietorship. There's no real reason to encumber yourself with all the reporting requirements of a corporation unless you're benefiting from tax implications or protection from liability."

Flexibility. Your goal is to maximize the flexibility of the ownership structure by considering the unique needs of the business as well as the personal needs of the owner or owners. Individual needs are a critical consideration. No two business situations will be the same, particularly when multiple owners are involved. No two people will have the same goals, concerns or personal financial situations.

Future needs. When you're first starting out in business, it's not uncommon to be "caught up in the moment." You're consumed with getting the business off the ground and usually aren't thinking of what the business might look like five or ten-let alone three-years down the road. What will happen to the business after you die? What if, after a few years, you decide to sell your part of a business partnership?

The issue of ownership was a key one for EnviroTech. "When we started EnviroTech," Kalish remembers, "our reasoning for forming it as a corporation was because of ownership; we wanted to be able to bring in stockholders as we grew."

"A corporation's capital," Baker says, "can be expanded at any time in a private offering by issuing and selling additional shares of stock. This is especially helpful when banks are being tight with money."

Another important question to ask yourself is, "What do I want to happen to the business when I'm no longer around to run it?" While a sole proprietorship or partnership may dissolve upon the death of its owner or owners, a corporation can be readily distributed to family members.

Keep in mind that the business structure you start out with may not meet your needs in years to come. Many sole proprietorships evolve into some other form of business-like a partnership or corporation-as the company grows and the needs of the owners change.

The tax aspects of a sole proprietorship are especially appealing because income and expenses from the business are included on your personal income tax return (Form 1040). Your profits and losses are first recorded on a tax form called Schedule C, which is filed along with your 1040. Then the "bottom-line amount" from Schedule C is transferred to your personal tax return. This aspect is especially attractive because business losses you suffer may offset income earned from other sources. As a sole proprietor, you must also file a Schedule SE with Form 1040. You use Schedule SE to calculate how much self-employment tax you owe.

In addition to paying annual self-employment taxes, you must also make quarterly estimated tax payments on your income. Currently, self-employed individuals with net earnings of $400 or more must make estimated tax payments to cover their tax liability. If your prior year's adjusted gross income is less than $150,000, your estimated tax payments must be at least 90 percent of your current year's tax liability or 100 percent of the prior year's liability, whichever is less.

The federal government permits you to pay estimated taxes in four equal amounts throughout the year on the 15th of April, June, September and January. With a sole proprietorship, your business earnings are taxed only once, unlike other business structures. Another big plus is that you have complete control of your business-you make all the decisions.

There are a few disadvantages to consider, however. Selecting the sole proprietorship business structure means you're personally liable for your company's liabilities. As a result, you're placing your own assets at risk, and they could be seized to satisfy a business debt or legal claim filed against you.

Raising money for a sole proprietorship can also be difficult. Banks and other financing sources are reluctant to make business loans to sole proprietorships. In most cases, you'll have to depend on your own financing sources, such as savings, home equity or family loans.

If your business will be owned and operated by several individuals, you'll want to take a look at structuring your business as a partnership. Partnerships come in two varieties: general partnerships and limited partnerships. In a general partnership, the partners manage the company and assume responsibility for the partnership's debts and other obligations.

A limited partnership has both general and limited partners. The general partners own and operate the business and assume liability for the partnership, while the limited partners serve as investors only; they have no control over the company and are not subject to the same liabilities as the general partners.

Unless you expect to have many passive investors, limited partnerships are generally not the best choice for a new business because of all the required filings and administrative complexities. If you have two or more partners who want to be actively involved, a general partnership would be much easier to form.

One of the major advantages of a partnership is the tax treatment it enjoys. A partnership doesn't pay tax on its income but "passes through" any profits or losses to the individual partners. At tax time, each partner files a Schedule K-1 form, which indicates his or her share of partnership income, deductions and tax credits. In addition, each partner is required to report profits from the partnership on his or her individual tax return. Even though the partnership pays no income tax, it must compute its income and report it on a separate informational return, Form 1065.

Personal liability is a major concern if you use a general partnership to structure your business. Similar to a sole proprietorship, general partners are personally liable for the partnership's obligations and debt.

In addition, each general partner can act on behalf of the partnership, take out loans and make business decisions that will affect and be binding on all the partners (if the general partnership agreement permits). Keep in mind that partnerships are more expensive to establish than sole proprietorships because they require more extensive legal and accounting services.

Protecting Business with a Partnership Agreement

Starting a business with a partner? It may be difficult to talk about problems during your honeymoon stage, but that's exactly when you should. A written partnership agreement helps guide you when questions arise.

According to W. Thurston Debnam Jr., a partner with Smith, Debnam, Narron, Wyche, Story & Myers LLP, a law firm in Raleigh, North Carolina, a partnership agreement should answer the following questions:

- If a partner becomes disabled, how long will he or she get a share of the profits? If a partner dies, what happens to that share? A good way to deal with this issue: life insurance on all partners.
- Can the partners have other outside partnership interests? In particular, can interest be in similar or competitive businesses?
- What is each partner's investment? Is one investing cash and the other energy? Do any of the partners own equipment that you'll use in the business, and does that fact deserve consideration as part of the start-up investment?
- What are the responsibilities and duties of each partner? Be specific about each partner's role in the day-to-day operations of the company.
- What will you do if one partner wants to withdraw? Typically, you'll set up a buyout agreement, but it's a very good idea to decide on the terms before the situation arises. You'll also want to include a noncompete covenant.
- Can a partner pledge his or her interest as collateral for a loan?
- Are additional contributions mandatory? If the business needs capital in the future, are partners required to make capital contributions?
- How will you restrict partnership-interest transfers? Can a partner transfer his or her ownership to anyone, or can you limit that transfer? This means the remaining partners won't find themselves in partnership with someone they object to. This is frequently used to protect the business in the event that one of the partners gets a divorce and his interest becomes a part of the divorce settlement.
- How will conflicts be resolved? Most often, an arbitrator is used.

Debnam recommends that every business partnership-regardless of the relationship of the individuals-begin with a written agreement. "It ensures that the partners have the same vision," he says.

But there's another reason for a partnership agreement. "Poorly drawn agreements keep litigation attorneys in business," Debnam notes. "The best reason to have a good agreement is to avoid the legal fees when you have a meltdown.

The bottom line? Don't take this very important decision lightly, and don't make a choice based on what somebody else has done. Carefully consider the unique needs of your business and its owners, and seek expert advice, before settling on a particular business format.

The simplest structure is the sole proprietorship, which usually involves just one individual who owns and operates the enterprise. If you intend to work alone, this may be the way to go.

Organizing Production Using Economic Principles

Agricultural industries, producers and producer organizations are often counseled to develop strategies or strategic alliances to address changing market and political environments. Over the next twenty years, production agriculture will experience fundamental changes, which, because of its rapidity and permanence, could surpass the tremendous changes that have occurred over the past fifty years. As the structure of agricultural production changes, so will the role and scope of agricultural producer organizations. Surviving organizations will be forced to fundamentally restructure their mission, goals and purpose.

Consequently, the application of strategic business management concepts will be increasingly more important for these groups over the next decade than at any previous time. The purpose of strategic business management is to build a strategic or competitive advantage over rival firms or organizations which can lead to long-term above-average returns for a firm in an industry. In general, successful companies employ one of three strategies: (a) a low-cost strategy, (b) a differentiation strategy, or (c) a focus strategy. Each of these strategies provides direction for firm-level decision-making and implicitly develops entry barriers to protect the developed competitive position. In addition, it is essential for a firm to consider strategies to defend its competitive position, lest it be overtaken by other firms who adopt similar market strategies.

The best strategy is ultimately a function of consumer demand and the product/service attributes, core competencies, and managerial skills of each company. However, the worst strategy is being "stuck-in-the-middle," that is, being unable to compete with others on the basis of cost, value, or market specificity. In any case, rivals may undercut prices, maintain market share, or become the supplier of choice whenever change occurs in an industry.

In addition, strategies must be refined as market conditions change. Over the next twenty years, farms and ranches will gravitate toward one of two production structures.

The first type of production structure will be similar to many current farms and ranches in that undifferentiated commodity products will continue to be produced.

Only low-cost producers will survive in this sector. A second category of producer will also evolve.

Farms in this category will produce differentiated, identity-preserved products that focus on certain product attributes and consumer demands. Strategic business management abilities will be especially critical for farms that gravitate toward identity-preserved production. Agricultural producer organizations have historically performed the role of providing a unified voice in relation to commodity programs and other agricultural policies and as a conduit for information among producers.

Trade liberalization, an increasingly global food system, the decoupling of commodity program benefits from production, and advances in biotechnology and information technology will alter the focus of agricultural producer organizations. Surviving organizations will be those who change their primary objective from lobbying for traditional commodity programs to providing resources and services needed by producers to cope with change and to expand profit opportunities. Such organizations will continue to provide valuable lobbying efforts with respect to a new range of issues, such as intellectual patent rights, trade liberalization negotiations, contract law, and environmental awareness. In addition, new roles for agricultural producer organizations will also develop.

These will include performing clearinghouse functions for biotechnology information, facilitating strategic alliances and farmer-owned cooperative ventures, and developing new educational programs designed to improve members' strategic and risk management capabilities with respect to specialty food and fiber production. Some producer organizations may provide risk transfer functions for members, serve as contracting agents to facilitate identity-preservation, and organize production contracts that ensure supply availability of specialty food and fiber products.

The combination of agricultural industrialization, trade liberalization, information technology, decoupled farm programs, environmental concerns and consumer demands for food quality, safety, convenience and nutrition will lead to unprecedented change in the agricultural production and the food and fiber processing and distribution sectors. Successful farm and ranch managers and commodity organizations are likely to be those who develop strategies which allow them to survive and prosper in this changing environment.

TRANSFORMATION OF AMERICAN AGRICULTURE

Over the second half of the twentieth century, agriculture in the United States underwent an extraordinary series of revolutionary changes affecting its technological, sociocultural, and ecological character. The transformation of American agriculture that began at mid-century is often described in neoclassical economic commentaries as a "productivity revolution" and is said to have been the result of the unleashing of "progress" through the application of modern science and technology to the problems and limitations of traditional agriculture.

Often left unsaid in the neoclassical tale of a glorious "green revolution" of highly productive "miracle crops" is the fact that the industrialization of agriculture also involved a corresponding concentration in the ownership of land, water, and other resources with grave social, economic, and environmental consequences.

Starting around mid-century, capitalist corporations took effective control of the agricultural means of production in the United States and this led to the end of America's "agrarian ideal" and the mass displacement of small-scale family farms, which virtually collapsed in a wave of bankruptcy between 1960 and 1990.

It is important not to hold a "romantic" view of the family farm as if it represents a perfect social institution that promotes balance with nature and equality among people across the locations of class, gender, and race. Guthman notes that there are significant problems with the small-scale family farm ideal ... [and] the agrarian imaginary is equally bound up with a sort of cultural conservatism and even with Christian fundamentalism.... [A]grarian populism [has] roots in conservative notions of an organic society Moreover, by failing to question the race and gender relations that enabled the family farm ... [the agrarian ideal] inherently glorifies them ... [it] also ultimately uphold[s] white privilege by ignoring the racial history of U.S. land policy

The point in highlighting the decline of the family farm is merely to identify a particularly significant structural change that occurred during the post-1950 transformation of U.S. agriculture.

The size in acreage of the new "factories in the fields" grew to as much as ten times larger than the traditional family farm. This shift in the ownership and scale of agriculture involved an alliance among industrial capitalist interests, the producers and purveyors of scientific knowledge in the universities, and the state sector for example, U.S. Department of Agriculture, USDA, and the Bureau of Reclamation. After 1950, the federal government provided massive crop subsidies, research support, and infrastructure development, like the construction of dams, reservoirs, and transportation systems, to support the expansion of agribusiness. The era of big dam construction in the western United States dates to shortly after 1902, when the Bureau of Reclamation was established. The triad of agribusiness-university-government bureaucracy jointly developed and sought to manage the process of scientific, technological, and political economic development of agriculture for the benefit of American corporations.

It is often overlooked that the industrialization or "modernization" of agriculture was also accompanied by a process of sociocultural transformation involving the "Latinoization" of rural America. The flip side of the decline of the family farm was the growth in the number of nonfamily workers in agriculture, which increased from 1.5 million in 1950 to 3.3 million in 1997.

This increase corresponds with the process over the past five decades through which farm work was essentially transformed from a system characterized by co-operative labor in mostly white farming families to a system based on the hiring of seasonal and temporary migratory workers by agribusiness farm operations. About 80 percent of non-family farm laborers today are of Mexican origin. They constitute the heart of a new transnational rural proletariat, created by the rise of an increasingly globalized corporate agribusiness, and employed across rural America from the Pacific Northwest to the Deep South.

The mainstream neoclassical narrative of the triumph and progress of American agriculture hides another story, one that chronicles persistent patterns of class, race, and gender inequality and exploitation, social and racial conflict, displacement from the land and forced migration, and ultimately the devastation of rural communities.

The neoclassical tale of modernity's progress obscures decades of environmental abuse and socioeconomic upheaval that accompanied the two modern scientific revolutions in American agriculture: the so-called "green" and "biotechnology" revolutions.

The scientific reorganization of agriculture by industrial capitalist corporations is a central force underlying the decline of family farms and rural communities, disruption of local food security, destruction of traditional forms of agriculture and environmental knowledge, threats to wildlife and its habitat, and a growing epidemic of hunger, malnutrition, and obesity in the "midst of plenty."

Finally, this radical transformation of agriculture has been confronted by the rise of social movements including struggles by Latina and Latino farm workers, farmers, urban workers and consumers, the elderly and low-income communities, and environmentalists in search of sustainable agriculture and environmental justice.

The history of agriculture in the United States during the twentieth-century involved a transition from a nation of independent farmers and craft workers to a nation of landless workers and urban consumers dependent on increasingly transnational corporate agribusiness producers, processors, and distributors of food.

According to the U.S. Bureau of Census, in 1900, over one out of every three Americans was part of the farm-based population (29.8 million out of a total population of 76.2 million). By1940, there were still 30.5 million individuals in the farm-based population out of a total population of 132.1 million or 23.1 percent. Rapid change followed after mid-century, and by 1970 only 9.7 million persons remained categorized as farm-based or roughly 5 percent of the total population of 203.3 million. The downward trend has continued, and in 1990 only 4.5 million persons were categorized as farm-based, or less than 2 percent of a total population of 248.7 million.

Radical Changes in Farms and Ranches

This transformation is further illustrated by data on the changing number and size (acreage) of farms and ranches. Again, according to the U.S. Bureau of Census, the decline in the number of farms between 1910 and 1940 was gradual with a loss of 100,000 farms (from 6.4 to 6.3 million). After 1950, however, the rate of decline accelerated so that only 2.9 million farms remained in 1970 and only 2.146

million in 1990. This decline seems to have stabilized recently, and in 2002 there was a slight increase to 2.158 million farms.

Some of this increase is undoubtedly due to the dramatic growth in the number of Latina and Latino owned farms, ranches, and orchards, an effect of the Latinoization of rural America. These data are indicative of the structural transformation of agriculture involving a shift from small and medium-sized family-based farms to industrial agribusinesses, which are corporate-owned factory farms in which only one type of crop is mass produced (monoculture) under the rules of "economies of scale."

The rise of mass production agriculture involved the concentration of capital in the form of fewer but larger agricultural operations: In 1935, there were only 89,000 farms in the U.S. larger than 1,000 acres or about 1.3 percent of all farms; still, these managed about 29 percent of all the farm land. By 1993, the number of farms larger than 1,000 acres had increased to make up 9 percent of all farms but these controlled nearly 65 percent of all productive farmland.

Concentration of Capital and the Globalization of Food Production

The period since 1950 has witnessed the emergence and widening influence of industrial monocultures controlled by transnational corporations such as Monsanto, ConAgra, Archer Daniels Midland (ADM), Iowa Beef Processors (IBP), Cargill, and others. The concentration of capital, industrial economies of scale, and uniform monoculture models of farming were the most significant developments in twentieth-century agriculture. This concentration is so advanced that by 1997 all major food products were controlled by groups composed of no more than four corporations. In livestock slaughtering and processing:

- Limore than 55 percent of the production of broilers (meat chickens) was controlled by Tyson Foods, Gold Kist, Perdue Farms, and ConAgra;
- 87 percent of beef production was controlled by IBP, ConAgra, Cargill, and Farmland Industries;
- 60 percent of pork production was controlled by Smithfield, IBP, ConAgra, and Cargill; and
- 73 percent of sheep production was controlled by ConAgra, Superior Packing, High Country, and Denver Lamb.

Similar extreme levels of industrial concentration were already evident in the milling and processing of all major commodity crop categories:

- 62 percent of flour milling was controlled by ADM, ConAgra, Cargill, and Cereal Food Processors;
- 76 percent of soybean processing was controlled by ADM, Cargill, Bunge, and Ag Processors;
- 57 percent of dry corn milling was controlled by Bunge, Illinois Cereal Mills, ADM, and ConAgra; and

- 74 percent of wet corn milling was controlled by ADM, Cargill, Tate and Lyle, and CPC.

The concentration of capital in agriculture has been extended through the process of globalization. This domination of agriculture by a handful of corporations has meant that local communities everywhere have increasingly lost the ability to control access, affordability, and safety of their food supplies.

The globalization of food production was evident in data from the late 1980s when agribusiness giants like Cargill, ADM, and Monsanto already operated in dozens of countries. Cargill—perhaps the most global of these firms in 1990—operated in forty-nine countries including Mexico, where it had six different operations.

Ecological Consequences of the Global-Industrial Agribusiness Model

There are well-documented ecological consequences resulting from the historical patterns of capital concentration, monocultures, and growing disconnection between local communities and their food sources. Between the 1950s and 1970s, a first wave of environmental problems developed out of excessive reliance on mechanized monocultures and their associated inputs of chemical fertilizers, pesticides, and herbicides. Since the publication of Rachel Carson's classic *Silent Spring* (1962), the use of pesticides and other agro-industrial chemicals has increased dramatically. For example, in California the use of pesticides increased between 1950 and 1999 by a factor of five accounting for $1 billion in agribusiness-related expenditures or about 12 percent of the U.S. total that year.

GREEN REVOLUTION

It was an effort by the United States to launch a modern industrial revolution in agriculture based on scientific research conducted for the development of high-yield varieties of food crops including maize, wheat and rice. The high yields of the new select miracle crops were only possible under a new regime of chemically-intensive agriculture. For this reason, critics note that the so-called miracle crops were not so much high-yield as high-input varieties.

Of course, low-income and subsistence farmers could ill-afford to make the substantial investments required to pursue this type of agriculture. Many of the first pesticides developed in the U.S. were derived from chemical weapons the military developed during Second World War. The technologies and practices of the so-called green revolution led to accelerated soil erosion and compaction, depletion of soil fertility, salinization and alkalinization of soils, pollution of surface and ground waters, loss of traditional farm land to infrastructure development, displacement of small subsistence-oriented farmers, destruction of wild life habitat, and loss of agricultural biodiversity.

In the 1950s, the United States exported its so-called green revolution to the rest of the world when Mexico became a major site for research on "improved" high-input varieties of hybrid wheat, corn, and other crops. The establishment in 1945 in Mexico of what became the CIMMYT, the Centro *Internacional para el*

Mejoramiento de Maíz y Trigo (International Center for the Improvement of Corn and Wheat), is an important chapter in the history of agricultural modernization and globalization. Less well understood is the role this model of industrial monocultures played in endangering the diversity of agricultural biodiversity and traditional farming communities in the U.S., Mexico, India, Brazil, the Philippines, and other countries that were the centers of green revolution research.

A second wave of scientific and technological change began in the late 1970s and early 1980s with the advent of research in plant genetics, which eventually resulted in the rise of a very large and globalized commercial agricultural biotechnology industry. The number of patented transgenic (genetically engineered) plants on the market has grown at an astonishing pace since 1990, and transgenic crops are now more than 50 to 60 percent of the critical commodity crops of soy, rapeseed, cotton, and maize grown in the United States.

The same corporations that were behind the miracle crops of the green revolution are now leading the biotechnology revolution and many of these are agrochemical interests that have recently absorbed smaller agricultural biotechnology companies. Instead of reducing the use of agro-industrial chemicals, transgenic crops have increased the use of herbicides and pesticides in almost all cases with increased hazards for farmers, farm workers, and wildlife. The rise of commercial agricultural biotechnology also represents a serious threat to the preservation of the diversity of agricultural crops by creating conditions for crop uniformity in rural landscapes. The simplification of cropping systems results in the loss of domesticated plant genetic resources because older varieties, most of them developed and preserved by small family farmers, become extinct.

Latinas and Latinos and Industrial Agribusiness

Since the mid-twentieth century, the historical experiences of Latinas and Latinos in the development of American agribusiness have been largely shaped by their status as temporary and seasonal migratory laborers. Government-sanctioned recruitment of Mexican workers for agriculture has a long history, dating back to the guest worker programs of the early to mid-twentieth century.

During World War I and II, the U.S. and Mexican governments agreed to a bilateral policy for the importation of massive numbers of Mexican workers, presumably to meet wartime labor shortages. The World War II policy was called the bracero program and resulted in the importation of several million workers between 1942 and 1964. The majority of braceros were destined for the agricultural sector although many worked on the railroads, mining, and even manufacturing.

Many Latinas and Latinos have experienced persistent poverty, hunger, and malnutrition in the midst of an agricultural revolution that increased the productivity of farming and the volume of food available for consumption. The availability of food for consumption (supply) has not translated into equitable access. The American agribusiness revolution did not lead to less hunger or to the elimination of malnutrition and related health problems.

On the contrary, the advent of agribusiness has resulted in wider disparities in access to nutritious foods and at the same time has undermined the viability of traditional foods and foodways.

As we have seen, the advent of industrialized agribusiness is a major storyline in American agricultural history. But the contributions of Latinas and Latinos to that history are more than a matter of a presumed and limited role as "unskilled" and "cheap labor" for growers. For Latina and Latino workers, a job in American corporate agriculture has meant earning less than a living wage, exposure to toxics and nominal environmental protection, lack of access to clean drinking water or toilets, lack of humane housing and medical care, and the underlying problem of the suppression of workers' right to organize and engage in collective bargaining.

The USDA also has a deep-rooted pattern of discrimination against farmers of color (African American, Latina and Latino, and Native American). Discriminatory treatment of farmers of color is part of the history of the land grant college-agricultural extension service complex and its research programs and development priorities, which have favored large-scale monoculture farming. Latina and Latino farmers have experienced inequitable access to private and public credit markets resulting in land loss due to predatory lending practices, real estate speculation, and rural gentrification.

This issue is of utmost importance, particularly because Latinas and Latinos represent one of the few ethnic/national-origin groups to have experienced an increase in the number of owner-operated farms, ranches, and orchards. According to official agricultural census data, between 1987 and 1997 the number of Latina and Latino owned or operated farms in the United States increased by nearly 40 percent. The increase was most salient in the Southwest and Northwest. For example, in the state of Washington, the number of Latina and Latino owned or operated farms including commercial orchards increased by 343 percent during that period. Significant growth of the Latina and Latino farm sector was also observed in Arizona, California, Colorado, Florida, New Mexico, Oregon, and Texas. Michigan, Wisconsin, Minnesota, and other Midwestern states also experienced growth. There are Latina and Latino farm owners and operators in every state in the nation. There were 16,183 farms operated by Latinas and Latinos in 1982; 17,476 in 1987; 20,956 in 1992; and an estimated 23,000 in 1997. At this rate, there will be at least 50,000 Latina and Latino operated farms by 2022.

Visions of Sustainable Production

A critical history of Latinas and Latinos in agriculture must acknowledge the loss of local food security caused by the globalization and concentration of agriculture. It must address the failure by the organic agriculture movement to integrate environmental justice principles into its visions of sustainable production. It must include discussion of the failure by the academic establishment to recognize and value the legitimacy of the local agroecological knowledge of farmers of color. Despite these scholarly shortcomings, Latina and Latino rural and urban communities are playing a significant role in social movement struggles seeking to make agriculture more sustainable and socially just.

The Latina and Latino environmental justice movement is developing a progressive vision of sustainable agriculture to affect changes in the direction of public policy discourses. Environmental justice scholars and activists are contributing to the analysis of food production and consumption as social justice problems in local, regional, national, and global contexts. They are demonstrating the role of racial and class discrimination in the experiences of people of color with land loss, farm labor exploitation, hunger, malnutrition, and health problems. Environmental justice frames the goals of sustainable agriculture according to the values and objectives of social justice, equity, and autonomy. The environmental justice framework for sustainable agriculture is pursuing the following types of objectives:

- Farmworker rights to organize to attain workplace health and safety and economic justice;
- Elimination of racial and class discrimination in federal and state agricultural policies and especially in farm loan, subsidy, and conservation programs;
- Survival and flourishing of family farms owned and operated by people of color; restoration of the lost lands of African, Native, Latina and Latino, and Asian American farmers and recovery of traditional customary systems of local management of natural resources and watersheds;
- Local food security to promote self-sufficiency, autonomy, and health of communities of color;
- Development of organic, community-supported, and alternative methods of farming to serve low-income families and the elderly in urban and rural communities of color;
- Protection of traditions of seed saving of heirloom crop varieties by farmers of color and resistance against the "patenting of life";
- Adoption of the "Precautionary Principle" and solidarity with farmers, seed savers, and food consumers of color opposed to the commercialization and environmental release of GEOs (genetically-engineered organisms) and the marketing of transgenic foods;
- Recognition of the contributions of farmers of color in the development of place-based agroecological knowledge;
- Recognition of the ecological, economic base, and social services and benefits provided by farmers of color as managers of millions of acres of farmland, open space, wildlife habitat, and diverse and healthy cultural landscapes.

These objectives represent a clear challenge to the dominant industrial agribusiness model of food production and are redefining the struggle for sustainable agriculture as a movement that integrates environmental sustainability with human rights and social justice.

STRATEGIC READJUSTMENT TO THE CHINESE ECONOMIC STRUCTURE

Work will be done to enhance Beijing's capacity of modern agricultural production on the basis what is designated as basic farmland. In keeping with the characteristics and relative superiority of locally available resources, modern agricultural undertakings of different types will be developed and so will be undertakings for processing agricultural products. This is meant to eventually build up metropolitan-type agricultural production and processing bases serving the needs of life and production in the city. Protection of the ecological environment and formation of an ecological shield for the municipality should be taken as the point of departure in planning the urban greenery buffer belt and ecological spheres on the plain. Work in this regard should be focused on developing forestry, orchards, animal farming, tourism and other green undertakings that meet ecological requirements.

In keeping with the orientation for development of Beijing's economy, work will be done during the tenth Five-year Plan period to develop, in a big way, a technology-intensive economy with high-tech industries taking the lead, a service economy with improved urban functions as the key indicator, a cultural economy backed by abundant cultural resources available in the city, and an economy opening to the world and participating in international competition. This is designed to develop an economic structure that meets the requirements of Beijing in performing its functions as the national capital and brings into full play Beijing's relative superiority in resources. Also expected is the kind of mechanisms that is adaptable and able to conduct dynamic self-readjustment in open market competitions.

By readjusting and optimizing the distribution and structure of production, we will ensure a great development of non-publicly owned economic sectors, speed up Beijing's effort to build up an information economy and an information society, significantly increase Beijing's overall competitiveness and development potential, and ensure a sustainable, fast and sound development of Beijing's national economy.

Role of Science & Technology

Aiming at a large increase in the overall competitiveness of the various sectors of production, attaching full importance to the role of science and technology progress and development of information technology in promoting the upgrading the structure of production, accelerating the building of a modern service industry, developing high-tech industries in a big way, making positive efforts to modernize agriculture and upgrading the three industries to eventually develop a structure of production that conforms to requirements of Beijing in functioning as the national capital and is adaptable to market conditions and highly competitive.

Accelerating the development of the modern service industry and promoting the optimization of the current structure of the industry: There is the need to accelerate the development of Beijing's superior resources and their transformation into factors of production and rationalize the distribution of resources through proper guidance. This is meant to expand the scope of development for the service sector. Also in need is work in a big way to develop modern service undertakings that are knowledge-intensive and based on Beijing's superior resources. Meanwhile, traditional service undertakings will be transformed and upgraded. Work in this regard is meant to raise the level of development for the entire service sector.

Development of the modern service industry will be accelerated. Priority will be given to development of information, intermediary and other knowledge-intensive service undertakings. Development and application of information technology will be taken as the basis for our endeavor to develop and use information resources.

There is the need to constantly open new areas for development of information service, with the focus on development of communications, networking, mass media, consulting and other service undertakings. Also in need is work to cultivate the market and standardize its operations through energetic effort to develop intermediary undertakings providing accounting, auditing, appraisal and assessment and law services. Development of the financial industry is to be accelerated. Efforts to enable the financial industry to become market-oriented and go international will be promoted, and work will be done to speed up innovation of the financial system and financial products. Positive efforts will be made to help the financial industry to expand the scope of its service and fresh impetus will be given to the development of banking, stock trading, fund, credit, insurance, financing lease and other modern financial operations.

In step with the upgrading of consumption demand, positive efforts are to be made to develop new services catering to the needs of urban and rural people in their life. In-bound and out-bound tourist markets are to be expanded and new tourism products developed, including conference and exhibition tours, metropolitan tours, study tours, holiday tours and science tours. Also to be done is work to accelerate the construction of municipal-grade holiday zones and other tourism projects, further standardize tourist market operations, improve the tourism environment in an all-round way, and constantly upgrade tourism services and their quality.

All this is meant to enable tourism to eventually become a pillar industry for Beijing. Development of cultural, recreational, sports, body building, medical, health care and other services will be encouraged and given proper guidance. Work is to be done to promote development of the real estate market, with focus of the work placed on housing construction.

Meanwhile, steps will be taken to rationalize the housing prices on the open market, so that the real estate industry will develop in a stable, sound manner. The property management system will be reformed, and work will be done to enhance the role of community services and promote the development of community service undertakings. Development of rural service undertakings will, too, be accelerated. Major efforts will be devoted to development of service undertakings that cater to needs of international and domestic markets or are foreign market-oriented and capable of stimulate the growth of relevant sectors. The scope of development for the modern service industry will be expanded, with priority given to development of service undertakings in the fields of education and specialists training, conferences and exhibitions, international trade and modern materials flow.

Transformation and upgrading of traditional services mean use of modern business and managerial methods and service techniques to transform and upgrade traditional undertakings including those of trade and commodity circulation, communication and transport and social services. Guidance will be given to commerce in effort to readjust its functional and networking structures and its business operations. Priority will be given to development of commercial undertakings that conform to the characteristics of Beijing as the national capital and are capable of stimulating the growth of relevant economic sectors and of service undertakings that help improve the quality of residents' life. Work will be done to standardize farm produce fairs, wholesales markets for farm and sideline products, second-hand goods markets, markets for distribution and trading of renewable resources and leasing markets.

Meanwhile, major efforts will be made to develop chain business, materials flow distribution and dispatch, e-commerce, commissioned distribution and sales and other new marketing and service methods to significantly enhance the level of modernization and competitiveness of the circulation sector. To eventually set up a modern, comprehensive system of communications and transport, we will strengthen construction of public transport networks, modernize materials flow bases, improve transport routes and stations, renovate transport equipment and management. In a notshell, we will enhance the level of development for the communications and transport industry.

Major efforts will be devoted to developing industries using high and new technologies to optimize and upgrade the structure of production. Overall optimization and upgrading will continue to be the focus of the work to readjust the structure of production. In line with the principle of òconcentrating on doing something while not doing something else, we will take implementation of the Capital 2-4-8 Major Innovations Program2 as the core and, by relying on science progress, we will strive to advance the technological innovation drive in an all-round way. Gigantic efforts will be made to develop industries using high and new technologies, and spread use of high and new technologies and technologies that are advanced while practical to transform. Meanwhile, work will be done to upgrade traditional industries and develop industries suiting urban market needs. No effort will be spared to basically accomplish the task of optimizing and upgrading Beijing's structure of production in five years.

Development of industries using high and new technologies will be accelerated so that a new pillar will be built for the national economy. Priority will be given to development of genetic engineering, biological chips, ultra-large integrated circuits, and nano and aerospace technologies. No effort will be spared to achieve breakthroughs in research & development of these technologies and put them to industrialized production, so that Beijing will be ahead of the nation in development of high-tech industries. The focus will be placed on development of five high-tech industries - industry of electronics and information, industry of bio-engineering and new pharmaceuticals, industry featuring an integration of optical, mechanical and electrical engineering, new materials industry, and industry for environmental production and multipurpose utilization of resources. This is aimed at developing a range of high-tech products of famous brands and with independent intellectual property rights, so that we can seize what is crucial to industrial development and effectively participate in international division of labor and competition. No time is to be lost to build ten high-tech industrial bases separately for development of software, micro-electronics, electronics and communications products, computers and networking products, nano materials, integration of optical, mechanical and electrical engineering, bio-engineering and new pharmaceuticals, fuel batteries, high resolution digital TV and liquid crystal displays. This is aimed at developing an industrial setup that is effective enough to ensure an intensive growth of high-tech industries. In the immediate future, construction will be accelerated on Badachu Science and Technology Park, Linhe Industrial Development Zone and Yizhuang Science and Technology Park, and work will be intensified to promote the development of the North China Microelectronics Base.

High and new technologies and technologies advanced while practical should be used as far as possible to transform and upgrade traditional industries. A good job should be done of ten projects designated for the purpose. These are, separately, for structural optimization and reorganization of the Capital Steel, expansion and renovation of the ethylene plant of the Yanshan Petrochemical Group, development of new types of jeeps and environmentally friendly cars, development of power plant fuel and firing system clearing processes, development of numerical control machine tools, upgrading of packaging and printing, development of energy-saving household appliances, development of functional building materials, development of environmental protection equipment for innocuous treatment, and industrialized production of fine chemicals.

The projects are meant to make Beijing's traditional industries more technology-intensive and enable these industries to produce new products of a higher grade to replace their old products. Positive efforts will be made to develop drinks, garments, packaging and printing, arts and crafts, tourism products, furniture and other industries serving needs of people's life and production. Industries and products that do not suit functions of Beijing as the national capital or are devoid of market prospects and competitiveness will be relocated or eliminated in an orderly way. Work will continue during the tenth Five-year Plan period to cut the surplus production capacity of some industries and restrict the production of conventional chemical materials and products, conventional synthetic fibers and rubber products and curtail the development of ferrous metallurgy and other obsolete industries. Metal mining, coking, chemical paper pulp and tannery industries and small, polluting factories will be eliminated by and large. Mines and factories that produce goods of inferior quality, cause waste of resources and heavy pollution and are not safe in production will be shut down in accordance with law. Work will also be intensified to eliminate obsolete equipment, technologies and technological processes.

The organizational restructure of enterprises will be optimized in line with the principle of concentrating on doing something while not doing something elseã%. Guidance and encouragement will be given to enterprises in conducting organizational restructuring either for integrated or multipurpose production, in keeping with the principle of specialized division of labor and requirements of economy on scale. In doing so, we will focus on developing a batch of companies and enterprise groups that are large in size, engage in trans-sectoral and trans-regional operations and are diverse in ownership. These are expected to strive for strengthened capital employment, enhanced development and innovation capabilities and improved marketing operations to promote development and upgrading of entire industries. Policy measures will be taken to encourage development of small- and medium-sized enterprises.

Assistance, support and guidance will be given to such enterprises—among them those that are technology-intensive in particular—in their effort to develop themselves for specialized production and production of products that are superb in quality, unique or new. The drive for stock right reform and realignment of assets will continue in small state-owned enterprises through realignment, joint operation, merger, contract operation and the cooperative share system or just by selling them out. Mechanisms that ensure withdrawal of enterprises from the market will be improved. Law will be invoked to declare bankruptcy those enterprises that have no market for their products and suffer from insolvency resulting from long standing deficits and inability to extricate themselves from the impasse.

Striving for better development of the building industry with a view to enhancing its market competitiveness. Work will be done to help construction enterprises enhance their technological and managerial level and improve the overall competence of their workforce. Major efforts will be devoted to developing domestic and international construction markets and strengthening work in contract construction and labor service outside China to expand the scope of development for Beijing's building industry and increase its market share. Management of design, construction, supervision and acceptance appraisal of construction projects will be improved to ensure quality of projects and a sound development of the building industry.

There is the need to standardize the construction markets and, under the principle of openness, fairness and honesty, work will be done to develop a system of open, competitive construction markets. Also needed is work to perfect the systems of administration and legislation over construction markets and the system of bid invitation and tendering for construction projects. Market supervision and administration will be strengthened. Behaviors of market subjects will be standardized. Intermediary service organizations and production factors markets will be developed to facilitate growth of the building industry. All this is aimed at creating the kind of environment and conditions for a sound development of the building industry.

Carrying out in an all-round way the policy of using science and technology to boost the development of agriculture and making positive efforts to speed up agricultural modernization and ensure sustainable development of agriculture.

During the tenth Five-year Plan period, work will be done to enhance the rural economic structural readjustment in line with the principle of enabling agriculture to become market-oriented, developing agricultural production by relying on science and technology, improving the profitability of agriculture and increasing incomes of the rural population. This is aimed at developing a suburban economy that is rational in structure, able to exercise its due functions and ecologically friendly.

Work will continue to see to it that the Party's basic rural policies are upheld and the results of their implementation are consolidated. The reform in rural taxation and fee collection will be advanced. Capital input will be increased through various channels for agriculture and rural areas to ensure that agricultural profitability and in incomes of the rural population will increase constantly and stability of rural social order can be maintained. To sum up, a foundation will be laid for Beijing's rural sector to lead the nation in basically modernizing agriculture.

Positive efforts will be made to readjust the structure of agricultural production. Major efforts will be devoted to development of an ecologically friendly agriculture and the projected shift of the farming sector from the dual pattern of production in which only food grain and cash crops are grown to a ternary pattern that features farming of food grain, cash and fodder crops. Development of ecologically friendly animal and fish farming will be accelerated, and the area sown to cash crops will be expanded.

Processing of agricultural and rural sideline products will be developed in a big way and repeated processing of these products will be encouraged for increased added value. Guidance and encouragement will be given to farmers in extending their operations of production in such a way as to include pre- and post-production services. Production of green food will be increased and production bases special for that will be set up. Work to prevent and control agricultural pollution will be strengthened to reduce the damage done to ecological environment by insecticides, plastic films for farm use and chemical fertilizers. The comprehensive production capacity of agriculture and its relative profitability are expected to grow through development of undertakings producing products of superb quality, crop seeds and breeding animals, greenhouse farming, export-oriented undertakings, and processing and tourism undertakings.

INNOVATION IN AGRICULTURD TECHNOLOGY

Major efforts will be devoted to innovation of agricultural sciences and technology. For that purpose, work will be done to improve the system of agricultural science and technology innovation. Success should be ensured in organizing implementation of programs to tackle key issues in agricultural science and technology development and of projects to industrialize agricultural production.

This is meant for major progress in developing modern bio-technology, technology for storage and processing of agricultural products, modern agricultural information technology and modern facilities and equipment for agricultural production. A range of modern, high-yielding agricultural parks and modern agricultural enterprises will be set up as part of the endeavor to promote industrialized high-tech agricultural production.

Work will be done to perfect the system of socialized services for agriculture. Enterprises processing agricultural products, rural wholesales markets and rural trade organizations, which play a pivotal role in developing a market-oriented agricultural economy, will be given still more assistance in their development. Organizations providing socialized services for agriculture is expected to develop in a big way, and farmers will be helped enter the market in a more organized way. Also needed is work to set up a modern market system and a system of quality standards for agricultural products, and perfect wholesales markets of origin. This is to ensure unimpeded circulation of agricultural products.

Encouraging Rural Enterprises

Rural enterprises will be helped in their revamping. Work in this regard includes promoting system renovation, assets realignment and adoption of the modern enterprise system by rural enterprises. Rural enterprises will be encouraged to work still more energetically in technological transformation as part of an effort to upgrade rural industrialization, promote development of higher-grade products to replace those proven obsolete, and stimulate the secondary and tertiary industries in rural areas. Priority will be given to development of industrial parks at the county- or district-level as well as township-level industrial zones, where under government guidance, rural enterprises will concentrate for still better development. This is also meant to raise the utilization rate of rural land resources and promote rural industrialization, urbanization and modernization to narrow, step by step, the differences between the city and the countryside.

Mountainous areas under Beijing's jurisdiction will be encouraged to develop, by proceeding from local conditions, economic undertakings special in nature. While properly protecting the ecological environment in these areas, we will help people there intensify their effort to achieve integrated development of locally available resources and develop special economic undertakings, so that they will get rich more quickly.

As always, priority will be given to construction of water conservation projects in these areas, which is the key part of the effort to alleviate poverty through development. Meanwhile, preferential policies in taxation, extension of credit loans and supply of electric power will be allowed to these areas in support of their construction and help improve living and production conditions there and narrow the differences between these and the plain areas. All sectors of society will be mobilized and organized to support the mountainous areas in their development.

Optimizing the distribution of different industries through readjustment.

This means proper guidance to ensure a rational distribution of resources and a stratified development of different economic sectors in light of the different functions designated to each part of the municipality, the characteristics of resources available to each and the relative superiority enjoyed by each. What is envisaged is a proper distribution of the various economic undertakings? tertiary undertakings that concentrate in the central part of the municipality, a high-tech industrial belt round the city proper, a modern agricultural zone in the plain area, and a modern processing industry zone beyond.

During the tenth Five-year Plan period, work will be intensified to rationalize the distribution of the various economic sectors and ensure a well-coordinated development of the economies of the different parts of the municipality. The objective will be achieved under the principle of facilitating exercise by the municipality of its due functions, enabling the different economic sectors to bring into play their relative superiority and ensuring rational utilization of resources.

The high-tech belt round the central part of the municipality. Areas along the highway round the urban sector of the municipality will be devoted to enterprises using high and new technologies with those in ZSTP and the five ZSTP industrial parks as the mainstay. The belt will serve as the base for intensive development of high-tech industries. Major efforts will be made simultaneously to develop knowledge-intensive service establishments such as those for science and technology development and education, as well as undertakings serving the needs of residents in their life. Still more effective planning and policy guidance will be provided to ensure a sound development of the various high-tech zones and parks by way of a relative division of labor. With development of ZSTP as the basis, plans will be made on a string of other science and technology parks and high-tech zones. These include the West Beijing Bio-pharmaceuticals Innovation Belt to specialize in research and development of bio-medicines and medical equipment, the South Beijing Aerospace Science and Technology Park for development of aerospace technology, the Xuanwu Xiannongtan Medical Science City for developing bio-medicines and making breakthroughs in medical research, and the Shunyi Genetics Science and Technology Park for research and application of genetic technology. Along the Badaling Freeway, Beijing-Miyun Highway and Beijing-Tianjin-Tanggu Freeway there will be a string of bases built specially for turning research results into commodities. This will provide high-tech industries with added scope for development.

Modern processing industry zones. These will be set up on the basis of the existing county- and district-level industrial parks and zones and those set up by key towns and townships. Enterprises in their respective areas are expected to concentrate in modern processing industry zones which, meanwhile, are to accommodate enterprises moved from the municipality's urban sector. These zones are to eventually develop into bases of modern manufacturing industry. Their distribution is to be based on a relative division of labor under municipal planning for functions each part of the municipality is designated to exercise and in light of the basic conditions there. Those in the northern and northeastern parts of the municipality are to concentrate on developing electronics, information, automobile, food, drinks, light and textile, garment and environmental protection industries.

Those in the southern and southeastern parts are to engage mainly in development of enterprises producing mechanical and electrical apparatus, pharmaceuticals and products featuring an integration of optical, mechanical and electrical engineering. Those in the western part of the municipality are required to concentrate on readjusting their structure of production and transforming the existing petrochemical, metallurgical and building materials enterprises. Guidance and assistance will be provided to encourage rural enterprises to relocate, for still better development, to county-and district-level industrial parks and zones and small rural towns that have sprung up in the course of urbanization. County- and district-level industrial parks and zones and industrial parks and zones in key towns and townships are expected to effectively develop their functions, do a good job of infrastructure construction and standardize their management and services. This is meant to create the kind of conditions favorable to concentration of rural industries for still better development and development of regional economies with distinct local characteristics.

Accelerating the process of relocating industrial enterprises in the municipality's urban sector and readjusting their structure of production. Still more vigorous efforts will be made in this regard. Industrial enterprises in areas surrounded by the Fourth Ring Road are obliged to move out in an orderly manner, by stages and in groups. Work will be done to see to it that in around five years, the proportion will drop from 8.74% now to 7% for the land space for industrial use to the total urban area. This is in part meant to enable the urban sector to attain the projected target of reducing the total amount of industrial pollutants in such a way as to basically get rid of the nuances that industrial enterprises cause the people to suffer.

Relocation of industrial enterprises in the urban sector and readjustment to their structure of production should be closely linked to the technological innovation drive, the strategic reorganization of the state-owned economic sector, use of foreign capital and construction of district- and county-level industrial parks and zones. This is meant to help the municipality's industrial establishment improve its economic performance and achieve a higher level of development, not just for a simple change of sites. Work will continue to effectively implement the various policy measures of support to prepare the kind of environment and conditions conducive to work in this regard.

Production and Inventory Management

Many agribusiness firms do not actively manage inventory. This does not mean that they ignore inventory. Rather, they hold large inventories because any potential savings from inventory reductions are far outweighed by the inventory-induced reductions in production, procurement, or transportation costs. Often economies of size cause long productions runs which lead to inventory accumulation. Simultaneously, seasonality leads to inventory buildups of key inputs like seed as well as outputs like corn. Economies in procurement such as forward buying in the food industry and quantity discounts increase inventories. Similarly, unit trains and other forms of bulk shipping discounts contribute to inventory buildups.

Many argue that the focus point (and perhaps the linchpin) of successful supply chain management is inventories and inventory control. So how do food and agribusiness companies manage their inventories? What factors drive inventory costs? When might it make sense to keep larger inventories? Why were food companies quicker to pursue inventory reduction strategies than agribusiness firms?

In 1992, some food manufacturers and grocers formed Efficient Consumer Response to shift their focus from controlling logistical costs to examining supply chains. Customer service also became a key competitive differentiation point for companies focused on value creation for end consumers. In such an environment, firms hold inventory for two main reasons, to reduce costs and to improve customer service. The motivation for each differs as firms balance the problem of having too much inventory (which can lead to high costs) versus having too little inventory which can lead to lost sales.

Supply Chain Management

A common perception and experience is that supply chain management leads to cost savings, largely through reductions in inventory. Inventory costs have fallen by about 60% since 1982, while transportation costs have fallen by 20%. Such cost savings have led many to pursue inventory-reduction strategies in the supply chain. To develop the most effective logistical strategy, a firm must understand the nature of product demand, inventory costs, and supply chain capabilities.

Firms use one of three general approaches to manage inventory. First, most retailers use an inventory control approach, monitoring inventory levels by item. Second, manufacturers are typically more concerned with production scheduling and use flow management to manage inventories. Third, a number of firms (for the most part those processing raw materials or in extractive industries) do not actively manage inventory.

Yet, such firms must be alert to changing conditions that may require more exact inventory management. One example would be if crops are marketed as small lots of value-added grain instead of commodities. Production proliferation in the seed industry may be another instance. Finally, whether due to food safety concerns, GMOs, food labeling, or the growth of organic food markets, identity preservation requires more precise inventory control.

Inventory Management

Inventory management is influenced by the nature of demand, including whether demand is derived or independent. A derived demand arises from the production of another product. For example, when John Deere knows its demand for a tractor, it can simply compute the demands for the parts, materials, and components needed to produce that tractor. Manufacturers of all sizes use such calculations which are part of flow management to manage inventories, schedule deliveries for inputs, and manage capacity. Flow management software has evolved from Materials Requirements Planning (or MRP) in the 1960s to the much more complex Enterprise Resource Planning (or ERP) of the 1990s. A flow management system is set in motion by the demand for end products.

Independent demand arises from demand for an end product. End products are found throughout a supply chain. Wheat is an end product for a grain elevator, as is flour for a miller or cereal for a grocer. By definition, an independent demand is uncertain, meaning that extra units or safety stock must be carried to guard against stockouts. Managing this uncertainty is the key to reducing inventory levels and meeting customer expectations. Supply chain coordination can decrease the uncertainty of intermediate product demand, thereby reducing inventory costs.

Customer Service and Inventory

The availability of inventory provides customer service. The Item Fill Rate (IFR) measures how often a particular product (often called a stock keeping unit or SKU) is available. A common metric of customer service, IFR is expressed as the percentage of time that a customer can obtain the item they seek. A firm may set its customer service order policy at 95%, seeking to fill 95% of the orders for an item from inventory.

However, life is a bit more complicated. A customer might not obtain what they seek for several reasons. The seller may have run out of a product due to an inaccurate forecast. Or the supplier may have shipped an incorrect package size or flavor. Products in inventory may be unfit for sale because of damage or an expired shelf life. Finally, a seller may not have the capability to accurately track inventory in their stores or distribution centers.

Safety Stock

To avoid shortfalls or stockouts, firms carry extra inventory known as safety stock. As more customer service is provided, a firm can expect sales to increase. However, as a firm tries to provide perfect customer service, logistical costs increase exponentially. Also, if a firm holds too much inventory, it can lead to low inventory turnover and hide operational problems. For example, carrying too much stock means that you might not discover that your supplier is frequently late with delivery times.

The structure of independent demand and logistical requirements vary by stage in the product life cycle (introduction, growth, maturity, and decline). During introduction, logistics must support the business plan for product launch, while preparing to handle potential rapid growth by quickly expanding distribution. At market maturity, the logistical emphasis shifts to become cost driven. In the decline stage, cash management, inventory control, and abandonment timing become critical. Over-abundance of products in the late maturity or decline stage will eventually result in obsolete products. The obvious difficulty is predicting how long each stage will last and how abruptly sales will fall in the decline stage.

The life cycle strategy typically involves getting to profitability quickly recuperating startup costs, then sustaining high profits for as long as possible, and finally acting decisively for products in decline to minimize losses. Understanding this life cycle can help managers select logistical tactics, inventory levels and supply chain designs. The ultimate goal for companies should be to have just enough inventory to satisfy consumer demand.

Another life cycle attribute is that demand uncertainty shifts as we progress through time. Product managers face substantial uncertainty during the introduction and growth stages, relative stability during maturity, and increasing uncertainty in decline. This uncertainty drives forecasting accuracy and the level of safety stock required to meet customer service expectations.

The coefficient of variation (CV) measures the stability of a product's demand, comparing the variability in demand to the size of the average demand. High demand variability in the introductory stage means it is difficult, if not impossible, to forecast demand. Thus, high levels of inventory must be held to meet even minimal customer service levels. In contrast, lower variability during maturity means that demand forecasts are quite accurate. However, inventory levels may still be large because they are based on larger sales volumes.

In addition to the vagaries associated with product life cycle stage, two other sources of uncertainty also drive the level of inventory. First, demand can vary from day to day, week to week, or seasonally. Second, there may be variability in lead time, or the time from when an order is placed until delivery is made.

Forecasting demand used to be more exact because products stayed in the mature product life cycle phase for a long time. Today many companies find it far more difficult to forecast sales because of product proliferation. Product line extensions result in more products that cannibalize sales and shorten the life cycle. Thus, more sales are coming from products in the erratic earlier stages of life, as opposed to sales from products in the mature stage of the life cycle.

Inventory Costs

Different models are used to manage inventory for products that are continually available like milk or products available for limited time like seed. The Economic

Order Quantity (EOQ) model determines the least cost level of inventory to carry, as well as costs. News Vendor models are used for products only available for a single period.

EOQ and News Vendor models have proved useful for managing inventory for many years, analyzing tradeoffs among major cost components. These models are robust and easy to customize to particular industries. Their approach to costing is similar reflecting levels of inventory, as well as shipping costs or quantity discounts.

Inventory costs fall into three classes: (a) carrying costs of regular inventory and safety stock; (b) ordering or setup costs; and (c) stockout costs. Inventory control systems balance the cost of carrying inventory against the costs associated with ordering or shortfalls.

First, carrying cost (or a cost to hold inventory) is comprised of capital costs, service costs, storage costs, and risk costs. A carrying cost involves the opportunity cost for holding inventory. If the firm did not have money tied up in inventory, it could either use the savings to make investments in other assets or pay down debt. Thus, a firm should first determine what it would do with any savings from a reduction in inventory. If the dollars are used to buy capital equipment, an appropriate opportunity cost is the firm's hurdle rate or its "required rate of return." If the dollars are used to pay down debt, the interest rate on the loan should be used to value the inventory. The other three aspects of carrying cost are non-capital costs.

The service costs are often masked in a firm's fixed costs. A firm should determine how much of its insurance and tax expense is associated with inventory. This is especially important in states that have an inventory tax. A firm has cash outlays for warehouses and materials handling equipment, either owning or leasing space from a distributor.

In either case, the firm should determine how much is spent on space. Inventory risk reflects characteristics of the product. Some items are more prone to be stolen, others are more likely to be damaged, yet others may become obsolete before a sale is made. In any case, risk means that if too much inventory is held, a certain proportion of the inventory will be unavailable for production or sale.

Inventory Reductions

To determine the cost of carrying inventory, one needs to know the average quantity of inventory, an inventory carrying cost (as a percent of product cost), and the average cost per unit of inventory. If a firm plans to use inventory reductions to fund other capital assets, inventory carrying cost might be 30% (25% for an opportunity cost and 5% for the service, space, and risk costs). If the firm plans to use the savings to reduce debt, the appropriate rate might be 12% (7% for the interest rate and 5% for the other costs). Regardless of the carrying cost rate being used, as a firm holds more inventory, carrying cost increases.

Firms carry extra inventory to guard against uncertain events. Known as safety stock, the purpose of this inventory is to provide protection against stockouts. Safety stock is costed just like regular inventory, it is an interest rate times the level of safety stock. The level of safety stock required to guard against a stockout depends upon the customer service level, the standard deviation of demand of the product, and lead time. Let's explain in greater detail.

Assume that it takes ten days from the time an order is placed until a shipment arrives and that on an average 20 cases are sold each day. Thus, over the teb days that we are waiting for the delivery (our lead time), we expect to sell 200 cases. If we trusted our forecast, supplier, and trucking company, we would simply hold 200 cases for the ten days. But we realize that forecasts are inaccurate, some suppliers are unreliable, and shipping times vary. If less is sold than expected during the ten days or if the shipment arrives early, we will still have inventory on the ten day and no customer service problems are encountered. However, if sales are above expectations during the ten days or deliveries are late, we might run out (or stockout) of product.

Managing the uncertainty surrounding safety stock is the key to reducing inventory levels. But in today's competitive environment, it is difficult to lower safety stock requirements for two reasons. First, some buyers especially large retailers are requiring higher customer service levels, which raise safety stock levels. Second, the product mix for many firms includes more new products with the corresponding greater demand variability. Thus, most firms seeking to reduce safety stock can only do so by focusing on aggressively cutting lead times.

The second cost to consider is ordering costs. Ordering costs include a cost for transmitting the order, receiving the product and placing it into storage, inbound transportation, and processing the invoice. Recent advancements in information technology have lowered this cost by a factor of six for many industries. A manufacturer uses the cost of a production setup instead of an ordering cost.

Finally, stockout costs involve lost sales when no inventory is on hand. Such costs fall as inventory (and customer service) levels increase. The relationship between stockout costs and inventory depends upon the accuracy of the demand forecast and the ability of the firm to recognize and react to a change in demand. Stockout costs depend on how a customer reacts to a stockout, the frequency of stockouts, and the availability of substitute products. Stockout costs can be very high if a lack of substitute products means that a customer will switch suppliers. In contrast, if buyers simply substitute a different product, stockout costs may be inconsequential.

In practice, many firms do not assess stockout costs because different divisions of a firm cannot reach agreement on what is the cost of running out. Marketing may desire a very high stockout cost to force a penalty cost on running out. Operations or finance may resist this as it leads to inventory buildups.

Service level goals can differ by the value placed on stockouts and indirectly carrying costs. A high stockout valuation will result in higher inventories and higher service levels. One way to evaluate an inventory management policy is to choose a service level target. From this target, the inventory policy will determine the inventory requirements and associated costs of providing that level of service. A higher service level implies that more inventory will be held as safety stock. The tradeoff decision occurs at the point where the cost of carrying extra safety stock balances the stockout cost.

Inventory levels are affected by customer service expectations, demand uncertainty, and the flexibility of the supply chain. For products with relatively certain demand and a long product life, it should be relatively easy to maintain desirable customer service standards even as inventories are reduced. However, for products characterized by erratic demand, a short life cycle, or product proliferation, a more responsive supply chain and larger buffer inventories may be needed to meet a desired customer service level.

Consumers are demanding more customer service from firms throughout the supply chain. Firms with high customer service levels may gain a competitive advantage over those that do not have the supply chain capabilities in place or the ability to manage them.

Firms who understand their demand recognize stockout costs and carry appropriate levels of inventory are ultimately better able to effectively manage inventory and provide the desired service level to customers. As industrialization affects agribusiness and agriculture in general, the importance of customer service and competitiveness will become critical for firms and supply chains.

Capital Budgeting: Principle and Procedures

Capital budgeting is the planning process used to determine whether an organization's long term investments such as new machinery, replacement machinery, new plants, new products, and research development projects are worth pursuing. It is also know as investment appraisal. It is budget for major capital, or investment, expenditures.

Many formal methods are used in capital budgeting, including the techniques such as:

- Accounting rate of return
- Net present value
- Profitability index
- Internal rate of return
- Modified internal rate of return
- Equivalent annuity

These methods use the incremental cash flows from each potential investment, or project. Techniques based on accounting earnings and accounting rules are sometimes used - though economists consider this to be improper - such as the accounting rate of return, and "return on investment." Simplified and hybrid methods are used as well, such as payback period and discounted payback period.

Net present value

Each potential project's value should be estimated using a discounted cash flow (DCF) valuation, to find its net present value (NPV). This valuation requires estimating the size and timing of all the incremental cash flows from the project. These future cash flows are then discounted to determine their present value. These present values are then summed, to get the NPV. See also Time value of money. The NPV decision rule is to accept all positive NPV projects in an unconstrained environment, or if projects are mutually exclusive, accept the one with the highest NPV(GE).

The net present value is greatly affected by the discount rate, so selecting the proper rate - sometimes called thehurdle rate - is critical to making the right decision. The hurdle rate is the minimum acceptable return on an investment. It should reflect the riskiness of the investment, typically measured by the volatility of cash flows, and must take into account the financing mix. Managers may use models such as the CAPM or the APT to estimate a discount rate appropriate for each particular project, and use the weighted average cost of capital (WACC) to reflect the financing mix selected.

A common practice in choosing a discount rate for a project is to apply a WACC that applies to the entire firm, but a higher discount rate may be more appropriate when a project's risk is higher than the risk of the firm as a whole.

Equivalent annuity method

The equivalent annuity method expresses the NPV as an annualized cash flow by dividing it by the present value of the annuity factor. It is often used when assessing only the costs of specific projects that have the same cash inflows.

In this form it is known as the equivalent annual cost (EAC) method and is the cost per year of owning and operating an asset over its entire lifespan.

It is often used when comparing investment projects of unequal lifespans. For example if project A has an expected lifetime of seven years, and project B has an expected lifetime of 11 years it would be improper to simply compare the net present values (NPVs) of the two projects, unless the projects could not be repeated.

The use of the EAC method implies that the project will be replaced by an identical project. Alternatively the chain method can be used with the NPV method under the assumption that the projects will be replaced with the same cash flows each time. To compare projects of unequal length, say three years and four years, the projects are chained together, i.e. four repetitions of the three year project are compare to three repetitions of the four year project. The chain method and the EAC method give mathematically equivalent answers.

The assumption of the same cash flows for each link in the chain is essentially an assumption of zero inflation, so a real interest rate rather than a nominal interest rate is commonly used in the calculations.

Real Options Analysis

Real options analysis has become important since the 1970s as option pricing models have gotten more sophisticated. The discounted cash flow methods essentially value projects as if they were risky bonds, with the promised cash flows known. But managers will have many choices of how to increase future cash inflows, or to decrease future cash outflows. In other words, managers get to manage the projects - not simply accept or reject them. Real options analysis try to value the choices - the option value - that the managers will have in the future and adds these values to the NPV.

Ranked Projects

The real value of capital budgeting is to rank projects. Most organizations have many projects that could potentially be financially rewarding. Once it has been determined that a particular project has exceeded its hurdle, then it should be ranked against peer projects (e.g. - highest Profitability index to lowest Profitability index). The highest ranking projects should be implemented until the budgeted capital has been expended.

Corporate Bonds

When a corporation determines its capital budget, it must acquire said funds. Three methods are generally available to publicly traded corporations: corporate bonds, preferred stock, and common stock. The ideal mix of those funding sources is determined by the financial managers of the firm and is related to the amount of financial risk that corporation is willing to undertake. Corporate bonds entail the lowest financial risk and therefore generally have the lowest interest rate. Preferred stock have no financial risk but dividends, including all in arrears, must be paid to the preferred stockholders before any cash disbursements can be made to common stockholders; they generally have interest rates higher than those of corporate bonds. Finally, common stocks entail no financial risk but are the most expensive way to finance capital projects.

Use of Capital Formation in Agriculture

Capital formation is one of the basic factors for increasing production. This is all the more important in agriculture where we are faced with the task of increasing production to keep pace with the increase in population against the odds of the vagaries of monsoon. Judicious use of natural resources for sustainable production of agriculture, adoption of advanced technology and development of infrastructure for facilitating all agricultural activities, ensuring food security in the broader sense of making adequate nutritious food available and accessible to all and making agriculture a profitable commercial activity at par with other industries in the arena of global economy are the problems that can be successfully tackled only with a strong capital base.

This requires a close monitoring of the status of capital formation which in turn hinges on the nature of statistical system and quality of data available for measurement of capital formation.

At present, the official source of information on capital formation is the Central Statistical Organization who provide estimates of capital formation for the economy as a whole as well for the individual industrial sectors including agriculture, as part of compilation of National Accounts Statistics in accordance with the concepts and definitions contained in the System of National Accounts (SNA) of the United Nations.

Capital formation in SNA has been defined within the framework of the national accounts system. There are conceptual dilemmas and practical difficulties in adopting a broader definition of capital formation in SNA. These are clearly brought out dialogically by SNA-1993 in a separate section. However, for enhancing the utility and resourcefulness of national accounts in economic analysis, policy making and decision taking, SNA recommends compilation of detailed accounts for sub-sectors of the economy as well as satellite accounts wherein alternative concepts, definitions and classifications can be introduced.

Agricultural Development

This cannot be ensured by confining attention to the activities within the boundaries of agricultural fields. It should encompass activities fully or partially meant for agriculture such as production of fertilizers and pesticides, development of agricultural markets, rural roads and communication; augmentation of facilities for agricultural credit for small and marginal farmers, agricultural education, research and development of agricultural technology which are the main source of increasing production under the limited availability of natural resources.

For monitoring agricultural growth it is necessary to have a broader measure of agricultural capital formation that includes capital formation in all these activities, which can be called capital formation for agriculture in comparison with capital formation in agriculture being compiled and presented at present in the National Accounts Statistics. This report concentrates on issues involved in the compilation of capital formation for agriculture.

Agriculture, apart from crop production, is also broadened to include the allied activities, namely, animal husbandry, forestry and fishing. The resources for these activities are closely related. People often are engaged in more than one of these activities. Therefore, this report takes up agriculture and allied activities for consideration, and the term agriculture used in the report refers to agriculture allied activities mentioned above.

This study looks into the concept of capital formation and the limitation for adoption of the broader concept in national accounting. Next, the procedure adopted by CSO for compilation of capital formation in general and agricultural capital formation in particular is examined. Grouping of sectors and industries in the System of National Accounts -1993 is briefly touched upon to examine the possible regrouping of items of capital formation in agriculture.

The procedure for working out capital formation for agriculture and estimates for the years from 1980-81 are given next. Finally, suggestions are made in respect of compilation and presentation of capital formation for agriculture.

National Accounts Statistics

It is the authentic source for estimates of capital formation in different sectors of the economy including agriculture. Any review of the concepts and procedures for compilation of capital formation in agriculture is, therefore, requires a clear picture of the concepts and methodology involved in the existing procedure of estimating capital formation in NAS.

Capital formation takes place in the production units. It consists of additions, less disposals, to fixed assets and change in inventories. Additions to fixed assets, called fixed capital formation, are the assets produced as outputs from process of production that are themselves used repeatedly or continuously in other process of production for more than one year. Inventories consist of materials and supplies meant for intermediate input in production; work in progress; and finished goods and goods for resale.

The total fixed capital used in production loses its productive capacity in course of time due to wear and tear or obsolescence. In other words, fixed capital gets consumed in the process of production. The extent of loss of its productive potential is known as Consumption of Fixed Capital (CFC) which is to be compensated by acquisition of an equal amount of fixed capital in the current year. Fixed Capital Formation computed without netting for CFC is known as Gross Fixed Capital Formation (GFCF). The term Gross Capital Formation (GCF) refers to the sum of GFCF and change in inventories. GCF less CFC is known as Net Capital Formation (NCF).

Classification of Sectors and Activities

SNA-1993 explains the various ways of presenting National Accounts on the basis of the same basic principles of accounting. The central frame work of SNA-93 consists of Integrated Economic Accounts, Supply and Use Table, Three dimensional Analysis of Financial Transactions, Functional Analysis, and Population and Employment Tables. Social Accounting Matrices and Satellite Accounts are also elaborated in SNA-1993 for presentation of greater details in different formats which may deviate from the central frame work.

An overall view of the economy is obtainable from the Integrated Economic Accounts which provides, for institutional sectors, the full sequence of accounts relating to production, distribution of income, use of income, change in assets and liabilities, changes in net worth, stocks of assets and liabilities and net worth.

An institutional sector consists of institutional units which are economic entities capable, in their own right, of owning assets and incurring liabilities and engaging in economic activities and in transactions with other entities.

The five mutually exclusive and exhaustive institutional sectors into which SNA-1993 divides the economy are: Non-financial Corporations, Financial Corporations, General Government, Households and Non-Profit Institutions Serving Households (NPISH). These sectors can be sub-sectored to have a detailed analysis and understanding of the economy. The full sequence of accounts is possible for each institutional unit, and the sum of these individual accounts make up the sequence of accounts for the entire economy which is outlined in the flow chart at Annex II.

An institutional unit may be engaged in production or consumption or both. The unit engaged in production is called an enterprise. The non-financial corporations and financial corporations do not take part in final consumption. The units of the other three institutional sectors are engaged in both production and consumption.

An institutional unit may be engaged in more than one activity of production. It may consist of more than one establishment where an establishment is a production unit situated in a single location in which only a single (non-ancillary) productive activity is carried out or in which the principal productive activity accounts for most of the value added. For the purpose of understanding, we may assume that an establishment is engaged in a single activity though there are examples to the contrary.

A group of establishments engaged in the same or similar kinds of activity is called an industry. Therefore, an industry cuts across institutional units and , in some cases, institutional sectors. The classification of industry adopted by SNA-1993 is according to the International Standard Industrial Classification (ISIC). The National Industrial Classification (NIC) of India is also in consonance with ISIC. A particular item of asset of an institutional unit may be used in more than one establishment of the institutional unit.

Therefore, the capital account for an establishment and consequently for an industry is not always possible to compile. Agriculture is not an institutional sector. According to NIC, it is an industry. The establishments of this industry are spread across all institutional sectors excepting, perhaps, the Financial Corporation. Therefore, compilation of capital formation for agriculture as an industry involves difficulties.

However, there is necessity for compiling the sequence of accounts in agriculture to have a closer study of the agricultural economy. For this purpose, FAO has proposed the System of Economic Accounts for Food and Agriculture (SEAFA) wherein a sub-sector of agricultural households will be identified for compiling details of accounts including the capital account. This system has bee modeled on the accounting principles of SNA. SEAFA, if implemented, would throw more light on the agricultural economy needed by the policy makers.

Method of Compilation of Capital Formation in National Accounts in India

Gross Value Added is complied in NAS by industry and capital formation is compiled by assets and industry of use. The institutional sectors adopted at present are: Public Sector, Private Corporate Sector and Household Sector. The assets considered for compilation of capital formation are: assets created by construction activities, machinery and equipment and change in stock. Acquisition of valuables and cost associated with transfer of non-produced assets are not included in the compilation.

The estimates of capital formation by asset-based approach and industry-of-use approach do not tally. Further, the asset-based estimate for the total economy is reconciled in comparison with saving estimate assuming that the estimates of saving are more reliable than the estimates of capital formation.

Asset-based Approach

In the asset-based approach, capital formation due to construction is obtained through commodity flow approach, machinery and equipment by using industrial survey results, increment in livestock by using Livestock Census results. Change in Stock is estimated industry wise.

CONSTRUCTION

The Gross Fixed Capital Formation (GFCF) is the same as the value of output in the construction activity less the cost of repairs and maintenance. Value of construction output is estimated separately for pucca and kutcha constructions. The estimate of output value of pucca construction is the same as the output of the construction which is estimated by commodity flow approach. Here the value of output is obtained from the estimates of value of five input materials, namely, cement and cement products, iron and steel, timber and round wood, bricks and tiles, and permanent fixtures and fittings used in construction plus compensation of employees, interest and profit. The breakdown of value of construction by Public, Private Corporate and Household sectors are obtained through expenditure approach by collecting the data from budget documents, RBI results on studies of sample joint stock companies, and the results All India Debt and Investment Survey (AIDIS) in respect of Household sector.

Adjustments are made so that the total expenditure of the three sectors thus estimated tallies with the value of output of pucca construction estimated through commodity flow approach.

Value of output or capital formation due to *kutcha* construction is estimated purely by expenditure approach. In the case of public sector, the components involved are:

- Afforestation
- Reforestation
- Soil conservation
- Area Development
- Other construction such as bunding, field drains and kutcha bridges
- Irrigation

- Roads and buildings
- Fifty percent of other construction in forestry

In the case of private corporate sector, *kutcha* construction covers expenditures on tea, coffee and rubber plantations. The Household Sector kutcha construction estimates are made using the results of AIDIS conducted once in ten years.

MACHINERY & EQUIPMENT

The value of output of machinery and equipment are obtained from Annual Survey of Industries (ASI) in respect of the registered sector and the proportion of value of output of machinery and equipment to the total output from ASI results is applied to the output of unregistered sector to get the value of output of machinery and equipment in the unregistered sector. Adjustments for export and import are also made and the value of machinery and equipment is obtained at purchaser's price by applying trade and transport margin.

Increment to Livestock

The estimate of value of net increases in the productive animals used as draught animals and those yielding milk, wool etc. (which gives NFCF) , and the value of other stock (which gives Change in Stock) are estimated using the results of Livestock Censuses conducted once five years. The census year results are extrapolated to get the estimates for the current year.

Industry-of-Use Approach

The GFCF for individual industry is obtained by estimating the GFCF for Public, Private Corporate and Household Sectors separately which are obtained from budget documents and the results of ASI, various sample surveys and by applying direct/indirect indicators.

Procedure for Compilation of GFCF in Agriculture Public Sector

Gross Fixed Capital Formation (GFCF) in the Public Sector is mainly due to irrigation projects undertaken by the Departmental Commercial Undertakings. There is a minor contribution by the Non-Departmental Commercial Undertaking on account of development of irrigation, horticulture, livestock and development of State Farms. Expenditure made by the Ministry of Agriculture, Rural Development etc. on crop Husbandry, soil and water conservation, preservation of wild life and other agricultural programmes leading to tangible or non-tangible assets, is not accounted as capital formation in agriculture, but included as capital formation under public administration.

Private Corporate Sector

Capital formation in the private corporate sector generated mainly due to plantation activities is estimated by collecting the data from the tea, coffee and rubber boards etc.

Household Sector

In the household sector, capital formation is due to construction activities such as digging of wells/tube-wells, construction of bunds and farm houses etc. which is estimated by using results of the All India Debt and Investment Survey (AIDIS) conducted once in ten years. For the post survey years the estimates are made by projecting the base year results by using indices of rural construction and agricultural production specially computed for the purpose. Acquisition of agricultural machinery and transport equipments are estimated by extrapolating AIDIS results by using the results of Annual Survey of Industries.

Increment in Livestock is estimated by extrapolating the results of Livestock Censuses conducted once in five years. As regards forestry, most of the forests are owned by the government and the capital formation is compiled from the budget documents. For the fishing activity , GFCF is estimated as net addition to capital stock comprising mechanised & non-mechanised fishing boats, fishing gears etc., by using the results of Indian Livestock Census (ILC). For the post census years, the results are extrapolated to get the estimates.

The components of GFCF in Agriculture & Animal Husbandry in the base year of NAS, namely, 1993-94.

It is seen that the growth rate of capital formation in the Household Sector in the post AIDIS years is the growth rate of the indicators. The procedure for computing the indicators may be made explicit and the component-wise estimates be made available to the Ministry of Agriculture for a better understanding of the trend of capital formation. Similarly, Livestock Census conducted once in five years is the source of data for estimating capital formation due to increment in livestock.

However, the census is not conducted in the same year in all the States. In the absence of annual data, CSO extrapolates the census figures by using the earlier inter-census growth rates, which may turn out to be incorrect due to conditions that drastically affect the growth pattern in the post-census years. The Livestock Census results needs to be released within a reasonable time period and procedures for providing realistic estimates of different livestock population for the post-census years be explored by the Department of Animal Husbandry, for use by CSO in their estimates.

In the above procedure, it is seen that the expenditures of the government on soil and water conservation etc. are not included under agricultural capital formation though the total picture on capital formation, even according to the existing procedure can be had only if we add the above expenditures to what is shown as capital formation in the Agriculture Sector. However, neither the institutional nor industrial classification as per the SNA procedures can capture the total at one place.

The economic and purpose classification of expenditures made by Administrative Departments are culled out from the budget documents, wherein we get the quantum of capital formation expenditure incurred on agriculture. Budget analysis for identifying capital expenditure from the Central Budget is done by the Department of Economic Affairs (DEA) regularly. But the classificatory procedures adopted by CSO and the DEA do not match exactly. As regards state budget analysis the Reserve Bank of India take up the exercise and bring out the publication, 'State Finances – A Study of Budgets'.

However it only re-groups the budget classification for reporting capital expenditure under different heads as opposed to the finer analysis done by CSO for NAS purposes.

There is necessity for these agencies to coordinate with one another to compile comparable aggregates for their mutual benefit. This would also reduce time and effort made in this direction and enhance the utility of these estimates.

Capital Formation as compiled by CSO is broadly in accordance with SNA and the definition and coverage of agricultural capital formation in SNA is constrained by the necessity for consistency and coherence within SNA. Solutions for policy and planning issues cannot be obtained merely from the confines of SNA. In fact SNA itself recognizes this shortcoming and recommends compilation of satellite accounts in harmony with SNA. In respect of Agricultural Sector, the information need for the managers of agriculture, extends beyond the production activity. The economic status of the people engaged in agriculture *vis-à-vis* other sectors, the availability of infrastructure for production and marketing, and infrastructure for production of various inputs and services such as education and research are also to be monitored and developed for a holistic growth of agriculture. Therefore, apart from the present series of capital formation compiled by CSO, we may have two other series of capital formation in agriculture obtained in the following way:

- Capital formation for agriculture obtained by regrouping the CSO estimates, and
- Capital formation for agriculture obtained by including capital formation in agricultural education , research etc.

The present exercise is devoted to the first step.

Regrouping of CSO estimates to get Capital Formation for Agriculture

Most of the industrial sectors contribute directly or indirectly to the development of agriculture. For example, fertilizer production is meant only for use in agriculture. Pesticide is also mostly used in agriculture. Supply of electricity in the rural areas is utilized in agricultural activities such as irrigation. Rural roads provide facilities for transport of agricultural commodities. Construction of godowns, cold storages, development of agricultural markets provide facilities for getting the monetary returns for the production.

A considerable proportion of the goods traffic on road and rails is o account of transporting agricultural commodities. The rural cooperative banks and commercial banks lend loans to the farmers to facilitate increase in agricultural production. Agricultural education and research is out and out meant for developing agriculture only. Increase in the capital formation in these activities boost up the growth of agriculture.

Therefore, calculation of capital formation for agriculture can be made by taking an appropriate proportion of capital formation in different sectors as available from NAS, and adding them together. This can be worked out for Public & Private Sectors combined, and for Public Sector separately by adopting the norms given below. The quantum of GFCF in each of the industrial sectors that qualifies for inclusion in the GFCF for agriculture can be obtained according to the following procedure.

1. Agriculture, Forestry & Fishing: The entire capital formation in the sector is for agriculture only and so the entire GFCF in these sectors qualify as GFCF for agriculture.
2. Mining & Quarrying: Capital formation in this sector is not meant for growth of agricultural production. Therefore, no GFCF of this sector is apportioned into agriculture GFCF.
3. Manufacturing: The activities of manufacturing fertilizers, pesticides and agriculture machinery produce goods meant for use in agricultural production. The entire capital formation in fertilizer and agriculture machinery industries are for agriculture. Pesticides are used in agriculture as well as in households. 59.4% of the pesticides is estimated to be used in agriculture on the basis of quantities of technical grade pesticides produced for agricultural and household uses. However, NAS gives the value of pesticides and fertilisers used in agriculture. It is assumed that the value of consumption of fertilizers is equal to the value of production of fertilizers and the entire fertilizer produced is consumed in agriculture. The value of production of pesticides is obtained by dividing the consumption value by 0.594. The total value of production of pesticides and fertilizers is obtained by adding these values of production. The proportion of the value of pesticides and fertilizers used in agriculture to the total value of production of pesticides and fertilizers is worked out, which turns out to be 96.16%. This proportion is applied on the GFCF of the fertiliser and pesticide industry obtained from Annual Survey of Industries (ASI) from 1980-81onwards to get the GFCF from the industry meant for agriculture for the corresponding years.

In the case of Public Sector, it is estimated on the basis of the share of Public Sector in fertilizer production which turns out to be 52%. On the basis of the information available from Pesticides Association of India, it is assumed that 90% of pesticides production is due to Private Sector. Weighted average of these proportions is computed by taking the corresponding consumption values as weights. The average works out to be 49.64% which is applied on GFCF for agriculture in fertiliser and pesticide at the aggregate level to arrive at GFCF for agriculture in the Public Sector. As regards Agricultural Machinery, the GFCF for individual years are taken from ASI data. It is assumed that the Public Sector's contribution to GFCF in agriculture machinery is negligible. Therefore, to work out GFCF for agriculture due to Public Sector, the GFCF of only the Fertiliser & Pesticide industries are included. The values at 1993-94 prices are obtained by using WPI for Industrial Machinery.

7

Accounting

Accounting is the language of business which are responsible for organizing all financial information into meaningful sets of data that can be used to conduct analysis of a business. Whether this analysis is done by managers (internal), owners (stockholders), or investors (external) some of the most important business decisions comes from this information. Financial Statement (as they are called) can be prepared yearly, quarterly, monthly or on demand of management or owners. The most useful analysis comes from documents that are prepared on a yearly basis because they contain the most information and give a better idea of how a business is performing over time.

BASIC ACCOUNTING DOCUMENTS

Balance Sheet

The Balance sheet is the first of the three main accounting documents. This document provides a single day snapshot for the business in three main headings. It is basically a statement of what is owned by the business (assets), who owns it (equity), and how much debt the business has (liabilities). The Assets must be equal to the combined value of both the liabilities and the owner equity.

Assets

Assets are things which can be owned such as anything that can be considered property. This includes both physical and non-physical objects that range from cars and printing presses to intellectual property such as patents and copyrights. One thing that won't be found here are things that cannot be owned, such as the employees and their associated skill sets. Assets are broken down into three main categories, current assets, fixed assets, and other assets.

Current Assets

(a) Current assets consist of cash and other things that turn into cash during the normal course of business.
(b) Examples include Cash, Inventory, and Accounts Receivable.

Fixed Assets

(a) Fixed assets consist of items that require an act of volition to convert into cash.

(b) Examples include Vehicles, Real Estate, Buildings, Tools, Computers and Furniture. While they are all worth something, they are not converted to cash on a day to day basis.

Other Assets

(a) This is the catchall for the rest of the assets.
(b) Examples include Patents, Copyrights, and Goodwill (Goodwill is not items given away for donation, it is the leftover amount paid for another business beyond the amount of its actual assets).

Liabilities

That which is owed, liabilities are effectively the bills and loans that the company currently owes. They are divided into two sections, current and other.

Current Liabilities

(a) Current Liabilities consist of things that require cash now or in the near future. If you borrow money from a friend for coffee this morning, and plan to repay it tomorrow morning, it's a current liability.
(b) Examples include Accounts Payable (bills received), Notes Payable (short term agreements where the total balance is due in the near future ex: water bill), Advance Payments (Retainers and other prepaid orders), and other accrued expenses.

Other Liabilities

(a) Other Liabilities is the residual category, if it is owed, and not currently due, it's here. If you borrow money from your parents for college, and they expect it paid back someday, it goes here.
(b) Examples include Vehicle Loans, Mortgages, credit cards, Student Loans, and Medical Bills.

Owner Equity

Owner Equity is the category for listing what stake the owner or owners of the company have invested in the company. Owner equity is a bit more confusing than the previous categories due to the split between owner types. There are effectively two different options here; the first is for small business, the other (significantly more complicated) for corporations.

Partnership and Sole Proprietorship Equity

(a) For these two business models, the Owner Equity section effectively consists of an account for each partner, and reflects the total amount invested in the business by that owner.
(b) Single line for sole proprietor, separate line for each partner
(c) If you go into this type of business model, have a written agreement!

Corporate Equity

1. For the corporation, Owner Equity is where we find the amount of money invested in the company by its various shareholders.
2. Examples include Preferred Stock (first chance at dividends and non-voting), Common Stock (Gets second chance at dividends), Excess Contributions (aka, Paid in Capital, Capital Surplus), and Retained Earnings (Accumulated earnings not released as dividends). Wikipedia has a nice description of Preferred Stock here and Common stock here.
3. Retained earnings are how you expand, and are owned by common shareholders
   (a) For a specific example, assume that at initial IPO, a company sells $100,000 of stock. The money earned on the sale of the stock is found under two headings, part under common stock, and part under excess contributions.
   (b) The first is listed under common stock and is the value for the number of shares sold at par value (10,000 shares, par = $1 per share, total $10,000).
   (c) The second part is listed under excess contributions and includes the rest of the amount of money spent for the stocks (10,000 shares, sold at $10 per share minus the par value of $1 per share, leaving a total here of $90,000).

A farm business large enough to adequately support a family is much too complex to manage from notes on a calendar or tablet. A detailed set of records is essential to making sound farm management decisions. This publication discusses the importance of farm records, explains the basics of bookkeeping, and outlines other major record keeping components and concepts including asset inventory, depreciation, profit and loss, enterprise accounting, and cash flow.

While computer software to do farm records is readily available, a manual system is discussed here to better illustrate concepts. Understanding a manual system will directly transfer to understanding a computer based system. Computer based record systems are widely available and should be considered when setting up a record system. Software capability to support farm records has grown dramatically in recent years. Different software packages differ in complexity and price.

However, the output—balance sheets, cash flow, income statements and enterprise accounts—provide the information necessary for farm business planning and management. Computers can be used to generate these documents; however, information on process and accounting is often absent.

KEEPING FARM RECORDS

There are a number of reasons for keeping farm records. First, farm records are a management tool. Farm records allow you to measure how efficiently you are using resources and to determine whether or not you are making any money. They help you define and evaluate success as measured by income generated for family living, retirement, and other needs and desires. Financial success is measured by profitability; if the farm business is not profitable it is not sustainable. Farm records are also essential for planning and decision making.

A second reason for keeping farm records is for income tax management. Good records simplify tax reporting and facilitate tax management to increase after-tax income. If you keep poor records, you may pay more taxes.

A third reason for keeping farm records is for obtaining credit. A good set of farm records allows you to determine credit needs and support loan requests. Properly kept records provide bankers financial information they need for making credit decisions, and good records also demonstrate your management ability.

Miscellaneous uses of farm records include pricing products for sale at a farmers' market, estimating the value of a CSA share, evaluating land leases, deciding whether to hire services or buy equipment, avoiding embarrassment from bounced checks, and evaluating farm insurance needs.

Characteristics of Good Record Keeping Systems

What are the characteristics of a good farm records system? Farm records should be easy to keep. They should provide essential information on a timely basis. They should contain an appropriate level of detail; complex farming operations with many and varied enterprises, such as multiple crops and livestock, require more detailed records. Other farms with few enterprises, perhaps a single crop, require less detail. A record keeping system should include:

- A business checking account to handle all business transactions.
- An income ledger to record all business income by calendar month.
- An expense ledger to record all business expenses by calendar month.
- An inventory that involves both the physical counting and valuation assignment.
- A depreciation schedule pro-rating the original costs of assets over more than one accounting period.
- A net worth statement or balance sheet summarizing assets and liabilities of the farm.
- An income or profit and loss statement that listsreceipts and expenses by type and the result is net profit or net loss.
- Cash flow statement measures the flow of funds into the business and the flow out of the business over the accounting period.

Enterprise records list receipts and expenses by enterprises.

In the following pages, each of these major components of a farm record keeping system will be discussed in detail. Additional components and, in particular, a cash flow statement can be an advantage depending on individual farm business needs.

Bookkeeping

Bookkeeping is the essential first step in organizing business transactions. Your books are simply a record of the money you spend and the money you earn. The information is written on a set of ledger sheets, which contain several columns, to keep track of where money goes and from what source it is earned. Categories can be used that correspond to those on form 1040F to simplify tax reporting.

Getting Started

A number of decisions need to be made about farm record systems. The desired level of detail and number of enterprise accounts need to be determined.

An enterprise account might be a group of similar crops like flowers, a single crop like broccoli, or a service provided to others. Determining the level of detail includes defining how many business enterprises to include under the farm business, deciding about home and living expense records, and about business interests outside the farm.

A decision needs to be made about the choice of accounting period. Should records be kept on a calendar basis or on a fiscal year? The time period selected should be the one that is the most suitable for the type of farm business. For example, some crop seasons end in midsummer, and this may be the best time to balance accounts. As a practical matter, many farm records are kept on a calendar basis to coincide with income tax reporting requirements. These and other record decisions will be discussed.

Business accounts often incur higher service charges than personal accounts. You may be able to open an "expense account" checkbook, which is just another regular account at a lower cost than a business account.

Bookkeeping Systems

There are two kinds of bookkeeping systems: single entry and double entry. Most farm bookkeeping does not require the refinements or the work of the "doubleentry" system. The "double-entry" method is a perfected system with built-in cross checks and automatic balancing. It requires two entries for every transaction:

This method is elaborate and usually requires a full semester of college to understand, plus it turns bookkeeping from an occasional nuisance into a full-time job.

This publication describes "single-entry" bookkeeping. Single-entry bookkeeping requires you tomake only one entry for every transaction, keeping paperwork and math to a minimum. This system will still provide you with the basic information you need to manage your farm and prepare your tax returns.

Accounting Methods

It refers to how you record transactions within the bookkeeping system. There are two accepted accounting methods: cash and accrual. Under cash accounting you record all taxable income, whether received in cash or as property, when it is received. With the accrual accounting, you record income when you earn it and expenses when you incur them, whether cash has changed hands or not.

Most farmers choose cash accounting. They find the record keeping easier and the greater flexibility of reporting income and expenses can be used to manage income taxes. Businesses which stock parts or keep inventories, such as retailers or manufacturers, are required to use accrual accounting because it provides a more accurate picture of business performance.

The study focuses on cash accounting; however, some accrual concepts for measuring business performance are described later.

Financial Management Skills Checklist

Low farm prices and low incomes, and the financial stress that results for many farm families reinforces the importance of financial management skills in having a successful farm business.

Financial management is more than a good accounting system and farm records– it includes careful use of borrowed funds and good capital investment decisions. Financial management also addresses the use of tax management strategies; the use of insurance to protect against financial losses that can arise from fire, loss of life, or health problems; the use of leases to gain control of assets; and the development of a sound estate plan.

USING ACCOUNTING INFORMATION FOR BUSINESS CONTROL AND PLANNING

Accounting Information Systems (AISs) combine the study and practice of accounting with the design, implementation, and monitoring of information systems. Such systems use modern information technology resources together with traditional accounting controls and methods to provide users the financial information necessary to manage their organizations.

Accounting Information Systems Technology

The input devices commonly associated with Accounting Information System include: standard personal computers or workstations running applications; scanning devices for standardized data entry; electronic communication devices for electronic data interchange (EDI) and e-commerce. In addition, many financial systems come "Web-enabled" to allow devices to connect to the World Wide Web (www) .

Basic processing is achieved through computer systems ranging from individual personal computers to large-scale enterprise servers. However, conceptually, the underlying processing model is still the "double-entry" accounting system initially introduced in the fifteenth century.

Output devices used include computer displays, impact and non-impact printers, and electronic communication devices for EDI and e-commerce. The output content may encompass almost any type of financial reports from budgets and tax reports to multinational financial statements.

MANAGEMENT INFORMATION SYSTEMS

Management information system are interactive human/machine systems that support decision making for users both in and out of traditional organizational boundaries. These systems are used to support an organization's daily operational activities; current and future tactical decisions; and overall strategic direction. MISs are made up of several major applications including, but not limited to, the financial and human resources systems.

Financial applications make up the heart of an AIS in practice. Modules commonly implemented include: general ledger, payables, procurement/purchasing, receivables, billing, inventory, assets, projects and budgeting.

Human resource applications make up another major part of modern information systems. Modules commonly integrated with the AIS include: human resources, benefits administration, pension administration, payroll, and time and labor reporting.

Multiple Uses

Accounting Information System cover all business functions from backbone accounting transaction processing systems to sophisticated financial management planning and processing systems.

Management accounting systems are used to allow organizational planning, monitoring, and control for a variety of activities. This allows managerial-level employees to have access to advanced reporting and statistical analysis. The systems can be used to gather information, to develop various scenarios, and to choose an optimal answer among alternative scenarios.

Financial reporting starts at the operational levels of the organization, where the transaction processing systems capture important business events such as normal production, purchasing, and selling activities. These events (transactions) are classified and summarized for internal decision making and for external financial reporting.

Cost accounting systems are used in manufacturing and service environments. These allow organizations to track the costs associated with the production of goods and/or performance of services. In addition, the AIS can provide advanced analyses for improved resource allocation and performance tracking.

Development

The development of an AIS includes five basic phases: planning, analysis, design, implementation, and support. The time period associated with each of these phases can be as short as a few weeks or as long as several years.

Planning—project management objectives and techniques The first phase of systems development is the planning of the project. This entails determination of the scope and objectives of the project, the definition of project responsibilities, control requirements, project phases, project budgets, and project deliverables.

The analysis phase is used to both determine and document the accounting and business processes used by the organization. Such processes are redesigned to take advantage of best practices or of the operating characteristics of modern system solutions.

Data analysis is a thorough review of the accounting information that is currently being collected by an organization. Current data are then compared to the data that the organization should be using for managerial purposes. This method is used primarily when designing accounting transaction processing systems.

Decision analysis is a thorough review of the decisions a manager is responsible for making. The primary decisions that managers are responsible for are identified on an individual basis. Then models are created to support the manager in gathering financial and related information to develop and design alternatives, and to make actionable choices. This method is valuable when decision support is the system's primary objective.

Process analysis is a thorough review of the organization's business processes. Organizational processes are identified and segmented into a series of events that either add or change data. These processes can then be modified or reengineered to improve the organization's operations in terms of lowering cost, improving service, improving quality, or improving management information. This method is appropriate when automation or reengineering is the system's primary objective.

The design phase takes the conceptual results of the analysis phase and develops detailed, specific designs that can be implemented in subsequent phases. It involves the detailed design of all inputs, processing, storage, and outputs of the proposed accounting system. Inputs may be defined using screen layout tools and application generators. Processing can be shown through the use of flowcharts or business process maps that define the system logic, operations, and work flow. Logical data storage designs are identified by modeling the relationships among the organization's resources, events, and agents through diagrams. Also, entity relationship diagram (ERD) modeling is used to document large-scale database relationships.

Output designs are documented through the use of a variety of reporting tools such as report writers, data extraction tools, query tools, and on-line analytical processing tools. In addition, all aspects of the design phase can be performed with software tool sets provided by specific software manufacturers.

Reporting is the driving force behind an AIS development. If the system analysis and design are successful, the reporting process provides the information that helps drive management decision making. Accounting systems make use of a variety of scheduled and on-demand reports. The reports can be tabular, showing data in a table or tables; graphic, using images to convey information in a picture format; or matrices, to show complex relationships in multiple dimensions.

There are numerous characteristics to consider when defining reporting requirements. The reports must be accessible through the system's interface. They should convey information in a proactive manner. They must be relevant. Accuracy must be maintained. Lastly, reports must meet the information processing (cognitive) style of the audience they are to inform.

Reports are of three basic types: A filter report that separates select data from a database, such as a monthly check register; a responsibility report to meet the needs of a specific user, such as a weekly sales report for a regional sales manager; a comparative report to show period differences, percentage breakdowns and variances between actual and budgeted expenditures. An example would be the financial statement analytics showing the expenses from the current year and prior year as a percentage of sales.

Screen designs and system interfaces are the primary data capture devices of AISs and are developed through a variety of tools. Storage is achieved through the use of normalized databases that assure functionality and flexibility.

Business process maps and flowcharts are used to document the operations of the systems. Modern AISs use specialized databases and processing designed specifically for accounting operations. This means that much of the base processing capabilities come delivered with the accounting or enterprise software.

The implementation phase consists of two primary parts: construction and delivery. Construction includes the selection of hardware, software and vendors for the implementation; building and testing the network communication systems; building and testing the databases; writing and testing the new program modifications; and installing and testing the total system from a technical standpoint. Delivery is the process of conducting final system and user acceptance testing; preparing the conversion plan; installing the production database; training the users; and converting all operations to the new system.

Tool sets are a variety of application development aids that are vendor-specific and used for customization of delivered systems. They allow the addition of fields and tables to the database, along with ability to create screen and other interfaces for data capture. In addition, they help set accessibility and security levels for adequate internal control within the accounting applications.

Security exists in several forms. Physical security of the system must be addressed. In typical AISs the equipment is located in a locked room with access granted only to technicians. Software access controls are set at several levels, depending on the size of the AIS. The first level of security occurs at the network level, which protects the organization's communication systems. Next is the operating system level security, which protects the computing environment. Then, database security is enabled to protect organizational data from theft, corruption, or other forms of damage. Lastly, application security is used to keep unauthorized persons from performing operations within the AIS.

Testing is performed at four levels. Stub or unit testing is used to insure the proper operation of individual modifications. Program testing involves the interaction between the individual modification and the program it enhances. System testing is used to determine that the program modifications work within the AIS as a whole.

Acceptance testing ensures that the modifications meet user expectations and that the entire AIS performs as designed.

Conversion entails the method used to change from an old AIS to a new AIS. There are several methods for achieving this goal. One is to run the new and old systems in parallel for a specified period. A second method is to directly cut over to the new system at a specified point. A third is to phase in the system, either by location or system function. A fourth is to pilot the new system at a specific site before converting the rest of the organization.

The support phase has two objectives. The first is to update and maintain the AIS. This includes fixing problems and updating the system for business and environmental changes. For example, changes in generally accepted accounting principles (GAAP) or tax laws might necessitate changes to conversion or reference tables used for financial reporting. The second objective of support is to continue development by continuously improving the business through adjustments to the AIS caused by business and environmental changes. These changes might result in future problems, new opportunities, or management or governmental directives requiring additional system modifications.

Internal Controls

Accounting information systems change the way internal controls are implemented and the type of audit trails that exist within a modern organization. The lack of traditional forensic evidence, such as paper, necessitates the involvement of accounting professionals in the design of such systems. Periodic involvement of public auditing firms can be used to make sure the AIS is in compliance with current internal control and financial reporting standards.

After implementation, the focus of attestation is the review and verification of system operation. This requires adherence to standards such as ISO 9000-3 for software design and development as well as standards for control of information technology.

Periodic functional business reviews should be conducted to be sure the AIS remains in compliance with the intended business functions. Quality standards dictate that this review should be done according to a periodic schedule.

ENTERPRISE RESOURCE PLANNING

These systems are large-scale information systems that impact an organization's AIS. These systems permeate all aspects of the organization and require technologies such as client/server and relational databases. Other system types that currently impact AISs are supply chain management (SCM) and customer relationship management (CRM).

Traditional AISs recorded financial information and produced financial statements on a periodic basis according to GAAP pronouncements. Modern ERP systems provide a broader view of organizational information, enabling the use of advanced accounting techniques, such as activity-based costing (ABC) and improved managerial reporting using a variety of analytical techniques.

Management accounting or managerial accounting is concerned with the provisions and use of accounting information to managers within organizations, to provide them with the basis to make informed business decisions that will allow them to be better equipped in their management and control functions.

In contrast to financial accountancy information, management accounting information is:

(a) forward-looking, instead of historical;
(b) model based with a degree of abstraction to support decision making generically, instead of case based;
(c) designed and intended for use by managers within the organization, instead of being intended for use by shareholders, creditors, and public regulators;
(d) usually confidential and used by management, instead of publicly reported; and
(e) computed by reference to the needs of managers, often using management information systems, instead of by reference to general financial accounting standards.

Traditional vs. innovative practices

In the late 1980s, accounting practitioners and educators were heavily criticized on the grounds that management accounting practices (and, even more so, the curriculum taught to accounting students) had changed little over the preceding 60 years, despite radical changes in the business environment. Professional accounting institutes, perhaps fearing that management accountants would increasingly be seen as superfluous in business organizations, subsequently devoted considerable resources to the development of a more innovative skills set for management accountants.

The distinction between 'traditional' and 'innovative' accounting practices can be illustrated by reference to cost control techniques. Cost accounting is a central method in management accounting, and traditionally, management accountants' principal technique was variance analysis, which is a systematic approach to the comparison of the actual and budgeted costs of the raw materials and labor used during a production period.

While some form of variance analysis is still used by most manufacturing firms, it nowadays tends to be used in conjunction with innovative techniques such as life cycle cost analysis and activity-based costing, which are designed with specific aspects of the modern business environment in mind. Lifecycle costing recognizes that managers' ability to influence the cost of manufacturing a product is at its greatest when the product is still at the design stage of its product lifecycle (i.e., before the design has been finalized and production commenced), since small changes to the product design may lead to significant savings in the cost of manufacturing the products.

Cause and Effect Accounting

Activity-based costing (ABC) recognizes that, in modern factories, most manufacturing costs are determined by the amount of 'activities' (e.g., the number of production runs per month, and the amount of production equipment idle time) and that the key to effective cost control is therefore optimizing the efficiency of these activities. Activity-based accounting is also known as cause and effect accounting.

Both lifecycle costing and activity-based costing recognize that, in the typical modern factory, the avoidance of disruptive events (such as machine breakdowns and quality control failures) is of far greater importance than (for example) reducing the costs of raw materials. Activity-based costing also deemphasizes direct labor as a cost driver and concentrates instead on activities that drive costs, such as the provision of a service or the production of a product component.

Dual Reporting Relationship

Consistent with other roles in today's corporation, management accountants have a dual reporting relationship. As a strategic partner and provider of decision based financial and operational information, management accountants are responsible for managing the business team and at the same time having to report relationships and responsibilities to the corporation's finance organization.

The activities management accountants provide inclusive of forecasting and planning, performing variance analysis, reviewing and monitoring costs inherent in the business are ones that have dual accountability to both finance and the business team. Examples of tasks where accountability may be more meaningful to the business management team vs. the corporate finance department are the development of new product costing, operations research, business driver metrics, sales management scorecarding, and client profitability analysis.

Conversely, the preparation of certain financial reports, reconciliations of the financial data to source systems, risk and regulatory reporting will be more useful to the corporate finance team as they are charged with aggregating certain financial information from all segments of the corporation.

In corporations that derive much of their profits from the information economy, such as banks, publishing houses, telecommunications companies and defence contractors, IT costs are a significant source of uncontrollable spending, which in size is often the greatest corporate cost after total compensation costs and property related costs. A function of management accounting in such organizations is to work closely with the IT department to provide IT Cost Transparency.

Given the above, one widely held view of the progression of the accounting and finance career path is that financial accounting is a stepping stone to management accounting. Consistent with the notion of value creation, management accountants help drive the success of the business while strict financial accounting is more of a compliance and historical endeavor.

An Alternative View

A very rarely expressed alternative view of management accounting is that it is neither a neutral or benign influence in organizations, rather a mechanism for management control through surveillance. This view locates management accounting specifically in the context of management control theory. Stated differently, management accounting information is the mechanism which can be used by managers as a vehicle for the overview of the whole internal structure of the organization to facilitate their control functions within an organization.

Planning and Financial Control

Planning and financial control are two of the most difficult and troublesome tasks faced by modern farmers and ranchers, but they are also the cornerstones of the management process.

Management usually is defined in three steps: planning, implementation, and control. This chapter focuses on how an accounting system can help you develop the information necessary for planning and controlling your farm business activities.

Approach of Accounting Systems

Before decisions can be made or analyzed, the information necessary for the decisions must be available. The primary goal of any farm or ranch accounting system should be to provide business management analysis and control. The accounting system should be geared toward the farm or ranch manager; if the accounting system is not used, it is for all practical purposes worthless.

Many uses of the accounting system relate to individuals other than the manager, so the system must be able to provide financial information for them too. The accounting system supports major management functions by providing the information necessary for making decisions. The accounting system should supply three types of information:

1. Scorekeeping, or evaluating performance (generally a retrospective look available in the financial statements);
2. Attention directing, to flag ongoing operating problems, inefficiencies, and opportunities (identified through analysis of the financial statements); and;
3. Problem solving, or analyzing the relative merits of alternative courses of action.

The accounting system provides information the farm manager needs for external reporting for tax and credit purposes; financial control of routine operations; business management analysis; and reporting to multiple owners for example, in corporations and partnerships.

Tax Requirements. The Internal Revenue Service (IRS) and most State income tax authorities require that enough business records be kept to justify all income and expense claims reported on an income tax return. The lack of standardized requirements for a minimum acceptable set of records has led some farmers and ranchers to store their cash register receipts, invoices, bank statements, and canceled checks in a box or file drawer and to do little more. Legally, such records are sufficient. This system can become an extremely expensive one, however, during an IRS examination, if the manager is called upon to substantiate claims made on an old tax return.

Other Taxes and Investments. If complete and accurate records are maintained, problems with estate, gift, and property taxes can be minimized. Furthermore, the ability to participate in investments outside of normal farm or ranch business activities can be enhanced by having the information readily available to determine whether a particular investment opportunity is financially feasible.

Credit Application. In recent years, lenders have come to stress repayment capacity of loans, as well as collateral security. Most borrowers now need to show that the investment for which the loan is intended will be able to generate enough income to pay back the interest and principal owed within the specified time period.

Financial Control of Routine Operations. Just as lenders are concerned with cash flows and repayment capacity, astute business managers have also become greatly concerned with cash-flow management. How much to borrow, either in long-term credit or in operating credit, is only half the story. When and how much to pay back is just as important as tight control of cash reserves. Paying operating money back after a sale may not be the wisest option if it puts the business in a cash-flow bind later. Interest charges must he analyzed in addition to liquidity needs of the business and of the family.

Preparing a realistic cash-flow budget is one of the vital steps in the annual recordkeeping process. A cash-flow budget is a projection of anticipated cash receipts and cash expenditures, by category, for a future time period typically one year. Borrowing and repayment plans are included. Cash budgeting involves all the steps required in the whole farm or ranch planning process: marketing (including price projections for inputs as well as outputs), yield projections, and enterprise combinations.

Despite the difficulty of preparation, the cash-flow budget helps document managerial abilities and loan repayment capacities. Furthermore, the cash-flow budgeting process can be extended one more step to provide an extremely effective financial control device. If monitored monthly, or even quarterly, the cash-flow budget can indicate potential problems before they arise. This ability to foresee problems allows the manager to adjust before the fact rather than react afterward.

Business Management Analysis for Strategic Planning

If a farmer or rancher is disciplined enough to develop and maintain a records system to meet income tax reporting and credit application needs, then virtually all the needed information will be available to meet what is probably the most important goal of a farm or ranch records system: business management analysis. Good farm or ranch business managers know exactly what their variable and total costs of production are. They know whether they are meeting the goals of their marketing plans or their cash-flow budgets. They have analyzed their strengths and weaknesses, both in physical terms and financial terms. They know where their business has been where it is now, and where it is going.

Corporations and Partnerships. Multiple-owner forms of business organization require more detailed records because of more intricate tax reporting requirements, State corporation laws and additional documentation needs of lenders. Perhaps the most important need for more detailed records in partnerships and corporations comes from the likelihood of problems and potential conflicts among the individuals involved.

8

Human Resource

Human resource development is an important factor in capacity building and improving the overall efficiency of functionaries involved in implementation, monitoring, evaluation, research and extension programmes. Training is a major component of Human Resource Development (HRD). Systematic training, planning, management and its implementation by making best utilization of resources available within the country helps in bringing about desirable changes in knowledge and upgrade skills of extension functionaries associated with the process of agriculture development.

The training infrastructure has been created to meet out the training requirements of all levels of extension functionaries, farm youth and farmwomen. Looking into the importance of training in capacity building of extension experts and farmers, this scheme is selected for the strengthening of extension services and dissemination of agricultural technology to the farming community.

The public research extension systems have played a major role in increasing agricultural production. In order to achieve such objectives conduct of seminars/workshops, conduct of training programmes and field demonstration to selected local farmers, presentation of awards for agricultural activities by the farmers, printing and publishing of literature and procurement of extension aids have been included in this scheme.

The agricultural extension setup in the department at present is very weak and hence to overcome this, there is a need to assess the state training needs and to sensitize the officers of the department.

KVK Lakshadweep being the part and parcel of Department of Agriculture, is the nodal Institute for imparting training, conducting FLDs & OFTs. Provision has been made to assist KVK financially. Hence, a component financial support to KVK also proposed in the scheme.

LEADERSHIP

No area of personnel management is more important than the leadership ability of the farm or ranch owner(s) or top manager. Few operations can justify having a specialized chief executive officer (CEO) whose primary responsibility is business leadership. The normal owner manager or manager must be both a leader and an operations manager. Leaders make sure that employees are doing the right thing.

Leadership is about effectiveness and results. Managers make sure that things are done right. Management is about efficiency.

The business leader must provide the vision, face reality and communicate reality and direction to the business participants. Leaders identify with the customer's needs, insuring the farm or ranch is customer focused.

Important functions of leaders in personnel management are impressing upon employees that they are capable, and empowering them to maximize their human capabilities. This cannot be accomplished with the command-and-control management philosophy.

Role of Business Leaders

Leaders must generate and sustain trust to gain employees following. Business leaders must develop employees that can identify problems and assist them in finding solutions. Leaders must motivate employees to achieve excellence in everything they do. Leaders also need to be decisive and timely in decision making when dealing with employees. Loyalty that is built among employees will carry through to customers. For successful business leaders, experience, competency, and a commitment to life-long learning have never been more important.

For a farm or ranch with multiple personnel to be competitive, there is no more important activity than personnel management. Modern management style focuses on the participatory involvement of all employees. The old command-and-control management style allows little opportunity for employees to fulfill their human capability in the modern information, performance, and accountability work environment. Job descriptions, building team effort, effective communications, office personnel, continuing education and training, a written business plan and business leadership are all necessary areas of focus. The participatory management style can lead the business to enhanced competitiveness through its people – the most valuable resource for the lasting success of the business.

HUMAN RESOURCES LEADERSHIP DEVELOPMENT PROGRAM

In today's farm/ranch management environment, it is challenging to be the operations manager and be responsible for working through employees to see that everything is done right and effectively as possible, while also completing other objectives. Too many times managers spend time doing the jobs that someone can be hired to do at the minimum wage rate and letting critical marketing and financial management decisions go undone.

The Danaher Human Resources Leadership Development Program (HRLDP) provides an accelerated training program for college graduates who are at the top of their class and have proven the ability to excel in today's fast-paced environment. The program will provide graduates, pursuing a career in human Resources, the opportunity to gain experience in the various facets of this exciting field. It is designed to provide future Human Resource leaders that will fill critical management positions. The program will build on the technical skills of the participants by providing them exposure to experienced professionals through structured assignments at various manufacturing and division locations. Assignments will typically last one year but will vary based on several factors. Participation in the program will require geographic flexibility. Assignments could be in a plant, commercial sales and service, or business headquarters environment. Each rotation will include detailed training and experience. To be successful in Human Resources within Danaher, professionals must show competence in-house by leading and participating in kaizen events; leveraging various management tools and processes.

Team Efforts

Although most farms and ranches have too few employees to justify forming teams, they can benefit from owners, managers and employees working as a team. In the words of Katzenbach and Smith, the reason to use a team or group approach is that "teams outperform individuals acting alone especially when performance requires multiple skills, judgement and experience." The key to successful team efforts summarized from Katzenbach and Smith are the following:

1. Group agreement on a common purpose – established urgency and direction.
2. Measurable performance goals established.
3. A defined working approach to accomplish tasks is followed.
4. Holding each employee mutually accountable for results for all involved.

The team effort must have a commitment to a purpose and have a measurable performance goal. Participants need to be rewarded for achievements through the efforts of the team. The team leader is a facilitator and is not expected to have the answer – that's the purpose of the team – to be more effective than an individual. Even small teams in a farm or ranch environment can greatly facilitate participatory decision making and motivate participants to a higher level of achievement.

The participants will be placed in three structured assignments: recruiting; associate relations; and compensation and benefits. Based on the participant's development plan, there is the potential for a fourth assignment, industrial relations. Selection of the assignment will involve the participant's preference, availability of the assignment, and the needs of the organization. The specific responsibilities within each assignment may vary based upon the immediate needs of the business. Some assignments share some of the same objectives. These objectives overlap from various assignments, as they are the foundation of sound human resources' practices.

Recruiting: Danaher strongly believes in promotion from within, when the talent and experience is available. It is their goal to fill 75% of all senior. management positions internally. However, as their organization continues to grow at an aggressive pace, it becomes necessary to recruit leaders from outside the company at all levels of the organization. Sourcing top talent is a primary focus of Human Resources and has a direct impact on their's financial success. Recruitment will be a part of each rotation. However, the Recruiting rotation is designed to provide an immersion into their talent acquisition process. Typical recruiting assignments include:

- Sourcing candidates for exempt level managerial positions.
- Managing outside recruiting sources.
- Identifying new selection sources.
- Coordinating the interview processes.
- Representing at career fairs.
- Conducting on-campus interviews.
- Developing position descriptions and job specifications.
- Developing and delivering job offers.
- Managing the onboarding and immersion processes.

Associate Relations: The majority of Danaher's manufacturing facilities are union-free. They work very hard at maintaining our union-free status by providing our associates the support and resources they need to have a successful and rewarding career. Typical associate relations assignments includes:

- Handling day-to-day associate issues.
- Coordinating associate events.
- Preparing and administering internal communication programs.
- Policy and procedural development.
- Facilitating safety and health committees.
- Managing workers' compensation claims.
- Coordinating and conducting training and development programs.
- Investigating and reporting accidents.
- Recruiting production associates.
- Human Resources planning.
- Tracking metrics and developing and executing countermeasures.

Management Styles

The old command-and-control management style isfading and is being replaced by a participatory management approach involving everyone in defining objectives, decision making and accountability. Better trained personnel, performance measurement, improved information systems, and two-way communication facilitate the idea that all personnel need to be involved and responsible. The owners and managers need to take the responsibility for including employee participation rather than simply giving orders and then checking if the job was done. Walk around management and two-way communication show respect for the ideas of all employees and are part of successful personnel management. Implementation of participatory management also requires the sharing of written goals, strategies, financial and production performance information. Tying reward systems to strategies and improved performance is part of the participatory management style. Reducing the levels and number of commanders may also reduce costs and lead to improved employee satisfaction

A change in management style requires a major change in philosophy. New skills must be developed or a change in the people in the top of the organization may be required. The development of communication and performance evaluation systems is necessary to achieve a change in personnel management philosophy.

Management by objectives and rewarding performance is much more achievable if everyone is participating in defining goals and objectives and being accountable for both their own and collective efforts.

All efforts toward participatory personnel management begin with everyone knowing their job responsibilities, the goals and objectives of the business, and how to work as a team.

Compensation & Benefits: Providing competitive benefits and compensation structures for our associates is a critical role of Human Resources. Our programs are designed to meet the wide range of needs and to reward and recognize our associates for their hard work, dedication, and commitment to Danaher's success. Typical benefit and compensation assignments includes:

- Evaluating benefits of new and potential acquisitions.
- Auditing levels of existing benefits.
- Educating and enrolling employees in optional services.
- Assisting employees in medical plan utilization.
- Conducting wage and benefit surveys.
- Recommending wage actions for hourly associates.
- Preparing compensation recommendations for salaried associates.
- Participating in the annual salary planning process.
- Participating in executive incentive plan, stock option, sales incentive plan processes.
- Working with managers in executing on-time performance appraisals.

Industrial Relations: A few of Danaher's facilities are organized by labor unions. In a union environment, we strive to maintain a positive relationship with our associates and with the local union leadership. Maintaining this relationship enables each facility to continue to successfully adapt to the ever changing manufacturing environment driven by the Danaher Business System. This relationship is paramount in the success of a unionized facility and Human Resources must take the lead in establishing this positive and cooperative working relationship. Typical assignments in the area of industrial relations include:

- Exposure to the collective bargaining process.
- Handling the initial steps of the grievance process.
- Participation on the labor-management committee.
- Conducting associate surveys and developing action plans to drive improvement.

- Exposure to the arbitration process.
- Coordinating associate activities and events.
- Handling day-to-day associate issues.
- Assuming ownership of internal communication programs.
- Facilitating safety and health committees.

It shows their confident that participation in this leadership program will prepare the participants for a successful career within any human resources team.

Relocation: Exposure to the different environments and processes within Danaher is key to the growth and development of the HRLDP participants. In order to facilitate this, the participants may be asked to relocate at the completion of each rotation. The selection of the assignment location will involve the participant's preference, availability of the assignment, and the needs of the organization.

Relocation is handled in accordance with Danaher's Smart Sum Value relocation policy. Associates are discouraged from purchasing a primary residence while in the rotational program as relocation costs for the sale and purchase of a home are not covered by the Smart Sum relocation policy.

People, human capital, are an important resourcein making a farm or ranchbusiness more competitivein today's business environment. Participatorymanagement is a anagement style that providesopportunities to realizehuman capital potential.

This fact sheet summarizes a few ideas from themany books that arewritten on the subject ofpersonnel management in modern business management. These ideas provide insight on ways to attain afarm or ranch's business goals through its people.

Topics briefly reviewed include management style,leadership, and taking care of employees.

CHANGING MANAGEMENT STYLE

Production agriculture has been and is still dominated by the command-and-control management style.

Many employees find this style convenient. If things go wrong it's the manager's fault. They would rather not be accountable for their actions. On the other hand, as the ranch manager's responsibilities broaden, employees must be expected to do their own thinking. It does not make sense for managers to try to be everywhere and to make all decisions.

The leadership role is more important. Communication, participation, and performance analysis must replace the commandand-control style. Most people want to make a difference and performance needs to be quantified and measured when it happens.

Ownership must initiate a change in management style. They must motivate participation. They must begin to identify and employ workers that respond to the participatory style. Again, the communication of goals and sharing of financial and performance objectives becomes the motivation.

In many situations, the current commander that enforces the style must be replaced or retrained. A change in management style often needs to be facilitated by training and motivational efforts by an outside professional that does not have a vested interest in changing working relationships.

Modifying the reward system, enhancing communication, and providing support with an effective management information system can facilitate the management style transition. The sooner an organization makes the commitment to change, the sooner the benefits can be achieved.

Encouraging Employee Participation

Employee participation is enhanced by developing written job descriptions that clearly define responsibilities, the basis for measuring performance and the reward system that recognizes varying levels of achievement. A commitment toward continued training and sharing of training costs needs to be spelled out to all personnel as part of their job description. The expectations toward the team efforts should also be spelled out in the job description. This reinforces the philosophy that there is a mutual dependence required to accomplish individual job and overall business objectives.

Job descriptions are also helpful in identifying and hiring of employees, especially when developed through a participatory effort and closely tied to business strategies.

Taking Care of Employees

Employees cannot be neglected. Successful management begins with employee considerations. The best trained and self-motivated personnel often do not stay in agriculture. This reality places an even higher value on participatory management, continued training, employee reward systems, benefit packages and living conditions that make a ranch job as favorable as possible. The total employee "family" must be part of the package. Often, it's the spouse and children and their education that have to be part of the total package.

The high cost of maintaining employees means they have to be efficient and effective in their jobs. Training and reward systems cannot be neglected.

Selecting and Terminating Employees

Likely the most difficult task of personnel management is selecting and terminating employees. One has to remember when facing this reality that "it is not the employees or providers of services you terminate who makes your life miserable, it's the ones you don't."

Too often ranchers keep employees that do not meet the job requirements. Incompetent employees are tolerated because the task of letting them go and hiring and training new employees is perceived as being too difficult. The cost of an employee not doing their job is not only the cash outlay but the cost of opportunities lost or the recurring cost of correcting problems because the job is not being done correctly.

The whole hiring and dismissal process is more effectively accomplished if the job description, performance evaluation, and mission, and objectives are clearly communicated. Employees need feedback to know how they are doing and what needs to be changed. Both managers and employees need objectively measured performance results. Documented underachievement can clearly establish when termination is required. This information also facilitates selection of replacements.

The job description and performance analysis need to communicate the expectations that ongoing training is the responsibility of all employees. Many farm/ranch owners and managers could join together to encourage private and public offerings of continuing education and training opportunities. Managers, professionals, and other top employees of the farm or ranch must stimulate and encourage continuing education and training efforts.

HR Management

A Human Resource Management System (HRMS) or Human Resource Information System (HRIS), refers to the systems and processes at the intersection between human resource management (HRM) and information technology. It merges HRM as a discipline and in particular its basic HR activities and processes with the information technology field, whereas the programming of data processing systems evolved into standardized routines and packages of enterprise resource planning (ERP) software.

On the whole, these ERP systems have their origin on software that integrates information from different applications into one universal database. The linkage of its financial and human resource modules through one database is the most impor tant distinction to the individually and proprietary developed predecessors, which makes this software application both rigid and flexible.

Function of Human Resources Departments

It is generally administrative and common to all organizations. Organizations may have formalized selection, evaluation, and payroll processes. Efficient and effective management of "Human Capital" progressed to an increasingly imperative and complex process. The HR function consists of tracking existing employee data which traditionally includes personal histories, skills, capabilities, accomplishments and salary. To reduce the manual workload of these administrative activities, organizations began to electronically automate many of these processes by introducing specialized Human Resource Management Systems. HR executives rely on internal or external IT professionals to develop and maintain an integrated HRMS. Before theclient–server architecture evolved in the late 1980s, many HR automation processes were relegated to mainframe computers that could handle large amounts of data transactions.

In consequence of the high capital investment necessary to buy or program proprietary software, these internally-developed HRMS were limited to organizations that possessed a large amount of capital. The advent of client–server, Application

Service Provider, and Software as a Service SaaS or Human Resource Management Systems enabled increasingly higher administrative control of such systems. Currently Human Resource Management Systems encompass:

1. Payroll
2. Work Time
3. Appraisal Performance
4. Benefits Administration
5. HR management Information system
6. Recruiting/Learning Management Training System
7. Performance Record
8. Employee Self-Service

The payroll module automates the pay process by gathering data on employee time and attendance, calculating various deductions and taxes, and generating periodic pay cheques and employee tax reports. Data is generally fed from the human resources and time keeping modules to calculate automatic deposit and manual cheque writing capabilities. This module can encompass all employee-related transactions as well as integrate with existing financial management systems.

The work time module gathers standardized time and work related efforts. The most advanced modules provide broad flexibility in data collection methods, labor distribution capabilities and data analysis features. Cost analysis and efficiency metrics are the primary functions.

The benefits administration module provides a system for organizations to administer and track employee participation in benefits programs. These typically encompass insurance, compensation, profit sharing and retirement.

The HR management module is a component covering many other HR aspects from application to retirement. The system records basic demographic and address data, selection, training and development, capabilities and skills management, compensation planning records and other related activities.

Leading edge systems provide the ability to "read" applications and enter relevant data to applicable database fields, notify employers and provide position management and position control. Human resource management function involves the recruitment, placement, evaluation, compensation and development of the employees of an organization. Initially, businesses used computer-based information systems to:

(i) Produce pay checks and payroll reports;
(ii) Maintain personnel records; and
(iii) Pursue talent management.

Online recruiting has become one of the primary methods employed by HR departments to garner potential candidates for available positions within an organization. Talent Management systems typically encompass:

- Analyzing personnel usage within an organization;
- Identifying potential applicants;
- Recruiting through company-facing listings; and
- Recruiting through online recruiting sites or publications that market to both recruiters and applicants.

The significant cost incurred in maintaining an organized recruitment effort, cross-posting within and across general or industry-specific job boards and maintaining a competitive exposure of availabilities has given rise to the development of a dedicated Applicant Tracking System, or 'ATS', module.

The training module provides a system for organizations to administer and track employee training and development efforts. The system, normally called a Learning Management System if a stand alone product, allows HR to track education, qualifications and skills of the employees, as well as outlining what training courses, books, CDs, web based learning or materials are available to develop which skills.

Courses can then be offered in date specific sessions, with delegates and training resources being mapped and managed within the same system. Sophisticated LMS allow managers to approve training, budgets and calendars alongside performance management and appraisal metrics.

The Employee Self-Service module allows employees to query HR related data and perform some HR transactions over the system. Employees may query their attendance record from the system without asking the information from HR personnel. The module also lets supervisors approve O.T. requests from their subordinates through the system without overloading the task on HR department.

Many organizations have gone beyond the traditional functions and developed human resource management information systems, which support recruitment, selection, hiring, job placement, performance appraisals, employee benefit analysis, health, safety and security, while others integrate an outsourced Applicant Tracking System that encompasses a subset of the above.

Human Resource Management in India: 'Where From' and 'Where To?

Over many centuries India has absorbed managerial ideas and practices from around the world. Early records of trade, from 4500 BC to 300 BC, not only indicate international economic and political links, but also the ideas of social and public administration. The world's first management book, titled *'Arlhãshastra',* written three millennium before Christ, codified many aspects of human resource practices in ancient India. This treatise presented notions of the financial administration of the state, guiding principles for trade and commerce, as well as the management of people. These ideas were to be embedded in organisational thinking for centuries. Increasing trade, that included engagement with the Romans, led to widespread and systematic governance methods by 250 A.D. During the next 300 years, the first Indian empire, the Gupta Dynasty, encouraged the establishment of rules and regulations for managerial systems, and later from about 1000 A.D. Islam influenced many areas of trade and commerce.

A further powerful effect on the managerial history of India was to be provided by the British system of corporate organisation for two hundred years. Clearly, the socio cultural roots of Indian heritage are diverse and have been drawn from multiple sources including ideas brought from other parts of the old world. Interestingly, these ideas were essentially secular even when they originated from religious bases.

In the contemporary context, the Indian management mindscape continues to be influenced by the residual traces of ancient wisdom as it faces the complexities of global realities. One stream of holistic wisdom, identified as the Vedantic philosophy, pervades managerial behaviour at all levels of work organisations. This philosophical tradition has its roots in sacred texts from 2000 BC and it holds that human nature has a capacity for self transformation and attaining spiritual high ground while facing realities of day to day challenges. Such cultural based tradition and heritage can have a substantial impact on current managerial mindsets in terms of family bonding and mutuality of obligations.

The caste system, which was recorded in the writings of the Greek Ambassador Megasthenes in the third century B.C., is another significant feature of Indian social heritage that for centuries had impacted organisational architecture and managerial practices, and has now become the focus of critical attention in the social, political and legal agenda of the nation.

One of the most significant areas of values and cultural practices has been the caste system. Traditionally, the caste system maintained social or organisational balance. Brahmins (priests and teachers) were at the apex, *Kshatriya* (rulers and warriors), *Vaishya* (merchants and managers) and *Shudra* (artisans and workers) occupied the lower levels. Those outside the caste hierarchy were called 'untouchables'. Even decades ago, a typical public enterprise department could be dominated by people belonging to a particular caste.

Feelings associated with caste affairs influenced managers in areas like recruitment, promotion and work allocation. Indian institutions codified a list of lower castes and tribal communities called 'scheduled castes and scheduled tribes'. A strict quota system called, 'reservation' in achieving affirmative equity of castes, has been the eye of political storm in India in recent years. The central government has decreed 15 per cent of recruitment is to be reserved for scheduled castes, and a further seven and half per cent for scheduled tribes. In addition, a further 27 per cent has been decreed for other backward castes.

However, the liberalisation of markets and global linkages have created transformation of attitudes towards human resource (HR) policies and practices. Faced with the challenge of responding to the rationale of Western ideas of organisation in the changing social and economic scenario of Indian organisation, practitioners are increasingly taking a broader and reflective perspective of human resource management (HRM) in India.

This manuscript has three main parts. In the first part is provided an overview of important historical events and activity that has influenced contemporary managerial tenets, the second part of the manuscript describes the emerging contemporary Indian HRM practices and indicates some interesting challenges. Much of the second part is also summarised on four informative Figures. The concluding section, the third part of the manuscript, succinctly integrates the two preceding parts.

VALUE OF CONTEXT OF HRM IN INDIA

The managerial ideologies in Indian dates back at least four centuries. Arthãshastra written by the celebrated Indian scholar-practitioner Chanakya had three key areas of exploration, (a) public policy, (b) administration and utilisation of people, and (c) taxation and accounting principles. Parallel to such pragmatic formulations, a deep rooted value system, drawn from the early Aryan thinking, called vedanta, deeply influenced the societal and institutional values in India. Overall, Indian collective culture had an interesting individualistic core while the civilisational values of duty to family, group and society was always very important while vedantic ideas nurtured an inner private sphere of individualism.

There has been considerable interest in the notion that managerial values are a function of the behaviours of managers. England, Dhingra and Agarwal (1974) were early scholars who contended that managerial values were critical forces that shape organisational architecture. The relevance of managerial values in shaping modern organisational life is reflected in scholarly literature linking them to corporate culture, organisational commitment and job satisfaction, as well as institutional governance. Thus, understanding the source of these values and in particular societal work values which link the macro-micro relationships and in turn organisational practices had become a popular line of enquiry, and a great deal of evidence has been presented to support the importance of national culture in shaping managerial values.

One of the most widely read formulations of this literature is the seminal work of Hofstede (1980) who popularised the notion of clustering culture in generic dimensions such as power distribution, structuring, social orientation, and time horizons. In turn, these dimensions could be employed to explain relevant work attitudes, job incumbent behaviours and the working arrangements within organisational structures. Two of these dimensions were individualism and collectivism.

The traditional social ethos from the ancient roots, which was developed over centuries, underwent profound transformation during the British rule. Consequently, in the contemporary context multiple layers of values (core traditional values, individual managerial values, and situational values) have emerged. Though the societal values largely remain very much anchored in the ancient traditions they are increasingly reflecting corporate priorities and values of global linkages. But in the arena of globalisation where priorities of consumerism, technological education, mass media, foreign investment and trade union culture predominate, newer tensions are becoming evident. For instance, contemporary Indian multi national companies and global firms in India have started shifting their emphasis to human resources with their knowledge and experience as the central area of attention in extending new performance boundaries. Considerable research evidence attests to this trend with particular relevance to greenfield organisations with little or no historical baggages in their organisational culture.

Within Indian traditions the choice of individualistic or collectivistic behaviour depends on a number of culturally defined variables. The dynamics of these variables are underpinned through three key elements guiding Indian managerial mindscapes.

These three constructs are Desh (the location), Kaal (the timing), and Patra (the specific personalities involved). Sinha and Kunungo (1997) claim that the interaction of these three variables determines the guidelines for decisional cues. This managing or nurturing of the outer layer of collectivism in an inner private sphere of individualism in which demonstrates the behavioural anchors in Indian organisational life.

Human Resource Management

he diverse nature of the Agriculture Industry combined with its long-standing traditions presents a unique challenge to human resource management. The evolution of farming from being wholly owned and operated by the family has prompted competition between producers and other Industry sectors for qualified labour. Now more than ever farm operators must be able to recruit and retain a high performance workforce for the continued success of their business. An aging workforce, combined with changing consumer trends and environmental regulations, as well as more technically advanced farming equipment have fostered a growing need for skilled labour.

Human Resource Management responds to these changes, by providing farm managers with the necessary education and training to become better employers through adequate workforce management.

MANAGING PEOPLE ON THE FARM

As agricultural businesses expand to keep up with economies of size, it is natural that many more producers are going to find themselves in the dual roles of farmer and employer. For many individuals, this is a daunting possibility. Hiring your first employee is never an easy process, but it certainly isn't something that should be avoided at all costs-especially if it is a crucial aspect of your operation's future success.

This course gives you first-hand accounts of how applying the principles of sound labour management can avoid serious problems and how effective management of staff can produce tremendous benefits. The stories are true and they come from Canadian producers.

Managing people on your farm will enable you to:

- manage foreign workers - seasonal and harvest employment
- Identify key considerations when planning to hire family members
- Determine if your labour needs match your current workforce
- Identify the importance of different means of compensation and the impact of work environment on successful labour management
- understand your role as an employer and manager of human resources on the farm
- upgrade your current skills in human resource management and training
- establish better recruitment and dismissal policies, associating the link between attracting qualified workers and profitability
- establish a human resource management policy to ensure maximum productivity from a satisfied workforce (performance appraisal, managing yourself and others)

Managing People on Your Farm is based on the excellent book of the same name from the Canadian Farm Business Management Council. Recognizing that people have different learning styles, the ABeSchool, in partnership with the Council, worked together to create this online course as a valuable self-directed learning experience for managers and supervisors working in a farm business environment. Additional, material has been added to the original resource, as well as media-rich features that will make your learning experience an exciting one!

Farm Safety – Creating a Safety Culture in Your Farm Business

As responsible employers, you strive to create a culture of safety within the working environment that follows employment standards and government regulations. In the farm workplace it is always a challenge to create a culture of safety that keeps employees informed without overwhelming them with complex documentation. The focus of this course is to assist farmers /industry in adapting to regulations implemented in 2006. The materials in the Ontario Farm Safety Association's employer's kit are integrated throughout the course.

After completing this course, you will be able to:

- Identify the legislated requirements and related laws under the Occupational Health and Safety Act and how you can comply with the requirements.
- Explain the purpose of the Employment Standards Act, (ESA).
- Distinguish between issues covered by the Human Rights Code, the Employment Insurance Act and the Workers Safety and Insurance Act from those covered by the ESA.
- Articulate the principle behind the OHSA as it relates to farming operations.

MIDDLE MANAGEMENT IN AGRICULTURE

Middle managers play a key role in organizations. As "active agents at the frontier of control" they are responsible for smoothing the workflow, handling exceptions, overcoming unexpected problems, and reaching goals and objectives. They also manage relationships at the workplace and maintain a positive atmosphere.

Middle managers are particularly vital to the functioning of agricultural and agribusiness operations, which are often smaller and leaner organizations, with fewer management levels. Middle managers play additional roles in agricultural operations by promoting family business values while fostering employee retention and job satisfaction. Thus, middle managers gain added significance; however, many agricultural and agribusiness organizations have taken middle management's contributions for granted.

In a similar fashion, agribusiness researchers have all but ignored middle management. Except for an *ad hoc* study on supervisors in the San Joaquin Valley, middle management research is virtually absent from agricultural and agribusiness journals. A recent search for 'supervisor' and 'manager' in the Agricola database turned up no relevant citations. This lack of research is even more notable given the pivotal role of middle managers in agribusiness, since it implies a lack of theoretical insights to support managerial decision-making.

The absence of middle management research in the agricultural sector contrasts with other economic sectors in the U.S. and in Europe. Traditional research focuses on the coordinative and supervisory role of middle managers, as reflected in many human resource management (HRM) textbooks. This role, positioned between senior management and employees, often results in increased stress and role conflict. Redman, Wilkinson, and Snaper cite one manager as saying, "I find it difficult in some ways because I've seen the management side of things, and what I've got to do as a manager is to keep staff happy, when perhaps I'm not all that happy myself ..." (p. 110). Delbridge and Lowe conclude, "Supervisors still hold key, yet contradictory, positions [...] and, as with the conflict and resistance they police, their role must be explored rather than assumed or ignored" (p. 424). While there is research on middle management in different sectors, this research lacks a comprehensive model of how middle management functions are accomplished in managers' daily practice. Dopson and Stewart argue that there is "no comprehensive body of theoretical or empirical knowledge on the role, function and responsibilities of the middle manager."

Research Literature on Middle Management

Few studies of middle managers are available in the U.S. agricultural literature. Most notably, Billikopf interviewed 42 farm supervisors in the northern San Joaquin Valley in California in 1995, using a convenience sample. Included were 19 first-line supervisors (foreman, assistant barn supervisor, working herdsman, crew leader, and lead cowboy), 14 mid- to upper-level managers (supervisor, manager, herdsman, and barn supervisor), and nine farm employers (grower, dairy farmer, and farm labor contractor) in agricultural specializations such as vineyards, dairy, fruit, vegetable, livestock and agronomic operations. Overall, interviewees were highly satisfied with their jobs (4.5 on a 5-point scale). Yet, 88% of interviewees identified job stressors, with 69% of these involving people management issues, such as organizing and assigning jobs, counselling, disciplining, and terminating employees, or defending company policies.

Bitsch (2006) analyzes the job attitudes of supervisors and middle managers in the green industry, based on a set of case studies with sixteen supervisors in thirteen operations. Interviewees were less likely to emphasize negative aspects of their work than positive ones. Achievement, job security, supervision, and interpersonal relationships emerged as contributing primarily to job satisfaction.

Recognition, the work itself, organization and structure, compensation, and personal life were more ambiguous, contributing to both satisfaction and dissatisfaction. The only predominantly dissatisfying factor was the working conditions. This was mainly caused by the number of hours supervisors had to work in their operations and their lack of scheduling flexibility compared to non-supervisory employees.

European Research Literature on Middle Managers

For non-agricultural sectors, the key streams of the middle management literature can be divided into research on European organizations versus U.S. organizations; these research streams are quite different in terms of basic assumptions about therole of the middle manager in the organization as well as theoretical paradigms.

The European research stream is informed by a critical paradigm. European researchers are concerned with the plight of middle managers (Hallier and James), arguing that the outlook for their future is "profoundly pessimistic". A growing body of research analyses the challenges to professional identities of middle managers. Middle managers are seen as being squeezed both by structural or cultural changes, as well as by technological streamlining. Further, recent initiatives in large organizations (e.g., delayering or flattening of organizations, self-directed teams, total quality management, etc.) have resulted in particular problems for middle managers.

Hallier and James point out that organizational restructuring has increased the pressure on middle managers, and hence contributed to managerial identity problems, which is supported by Balogun's findings.

By contrast, Delbridge and Lowe state that the purported "death of the supervisor" in the research literature has been greatly exaggerated (p. 423). Still others, such as Ogbonna and Wilkinson in their study of U.K. grocery middle managers, argue that the data regarding the changing role of middle managers are ambiguous. Hales concludes after analyzing data on first-line managers in different industries, "Even those contemplating the 'end of management' envisage the demise of an organizational stratum, not the abandonment of management as a function: a world without managers is not a world that is not managed.

The key question is where, or with whom, the functions of management and supervision reside". Thus, while the functions of middle managers will not disappear from organizations, European researchers view middle managers as squeezed and saddled with added responsibilities.

MIDDLE MANAGEMENT HRM PRACTICES MODELS

Van der Weide and Wilderom also note the rarity of rich observational data about managerial behavior. In their study, they videotaped actual behavioral interactions of thirty highly effective managers from twenty Dutch organizations, and analyzed these videotapes. Their model is unique since it is based on in-situ observations of managers deemed highly effective. The model describes four categories of behaviors: steering behaviors; supporting behaviors; self-defending behaviors; and sounding behaviors.

These managers used three behaviors most often: providing direction (categorized as a steering behavior); verifying (categorized as a steering behavior); and providing positive feedback (categorized as a supporting behavior). They note that sounding behaviors and self-defending behaviors have rarely been reported in the leadership literature.

Compared to the research described in the prior section, the model that Van der Weide and Wilderom propose includes a wider range of behaviors. They conclude that these effective managers have what they termed a "rich repertoire of behaviors to draw upon" (p. 12), since they do not limit themselves to behaviors of a single category or rely solely on "positive" behaviors.

In another study, Hamlin asserts that there are universally effective managerial behaviors based on case studies of three U.K. public sector organizations using interviews and questionnaires. He also describes a continuum of criteria for managerial effectiveness, which range from positive (effective organization and planning or proactive management; participative and supportive leadership or proactive team leadership; empowerment and delegation; genuine concern for people or looks after the interests and development needs of staff; open and personal management approach or inclusive decision-making; communicates and consults widely or keeps people informed) to negative (shows lack of consideration or concern for staff or ineffective autocratic or dictatorial style of management; uncaring, self-serving management or undermining, depriving, and intimidating behavior; tolerance of poor performance and low standards or ignoring and avoidance; abdicating roles and responsibilities; resistant to new ideas and change or negative approach).

In contrast to Van der Weide and Wilderom, Hamlin suggests that managers need to exhibit positive criteria to be considered effective and will be considered ineffective if they exhibit negative criteria, implying a more conventional view of effective managerial behaviors.

Research Methods

Given the lack of research on middle management in agriculture, and Yukl's (1998) critique of many leadership studies as failing to collect rich and descriptive information, this study relies on a qualitative approach. In particular Parry et al., and also Hamlin call for more qualitative research into managerial and leadership behaviors and suggest using a grounded theory approach. Grounded theory is an inductive approach to developing theory in the social sciences, first proposed by Glaser and Strauss.

Grounded theory can be considered the master metaphor of qualitative research (Bitsch, 2005), used by numerous researchers in a variety of fields in many ifferent ways. For this study, its distinctive approach to data analysis, and in particular, the constant comparison method, is most relevant. Differing from the original rounded theory approach, mixed approaches have become common as qualitative research has grown more prominent in a variety of fields (Charmaz).

In the absence of a theory of middle management, research needs to start with empirically based variable development and theorizing based on data (see the discussion among Sutton and Staw; Weick; and DiMaggio in Administrative Science Quarterly on the difference between theory and theorizing). Therefore, a prerequisite to developing a theory of middle management practice is an interpretive description of middle managers' use of HRM practices. This exploratory research focuses on what specific practices they use (and do not use) and how they accomplish their functions.

Relating to grounded theory, this study will develop substantive theory rather than formal theory, and should be positioned with "grounded theorizing" approaches rather than purist grounded theory.

Basic Characteristics of the Middle Managers

Similar to previous research (explicitly by Delmestri and Walgenbach; Dopson and Neumann; Dopson and Stewart; Ogbonna and Wilkinson; Redman, Wilkinson, and Snaper; implicitly by Hamlin; Osterman), the organizations' designations of their middle managers were used for this study. While many researchers treat the definition of the terms 'middle manager' and 'supervisor' as unproblematic (e.g., Batt; Delbridge, Lowe, and Oliver; Hallier and James; Harrington and Williams; Sims; Van der Wilde and Wilderom; Zaccaro and Banks), in reality the terms can reflect numerous formal positions in an organizational hierarchy, job titles notwithstanding. According to Hales, supervision is "the proximal and immediate direction, monitoring and control of operational work."

Thus, supervision is integral to any managerial position with subordinates. In the request for an interview, the researcher asked to talk to "a supervisor, someone who manages others, is in charge of managing employees." In most cases, the designated interviewees did not include first-line supervisors, who were more likely to be included in the group of non-supervisory employees. However, this issue is problematic in seasonal agricultural operations, because the number of employees supervised varies by time of year, as well as by task. Therefore, a middle manager with no subordinates during the winter months might oversee a large department with 30 or more employees during the summer.

Middle Managers' Functions and HRM Practices

Agricultural middle managers are different from middle managers in other industries, and there are several possible reasons for this difference. In some organizations, they have assumed more authority for HRM decisions because their organizations are flatter.

Many organizations have never grown elaborate, bureaucratic structures and therefore, have no need for delayering. These issues can be examined by analyzing middle managers' organizational functions and decision-making authority. Also, middle managers could depend more on informal management practices than formal authority within agricultural organizations.

Some middle managers have little input into any of these HRM functions, whereas others have full decision-making authority. Based on the managers interviewed, there is no obvious connection between job or management experience and decision-making authority. As shown in the following tables, those middle managers with little input have developed informal practices for managing their subordinates effectively. Even more surprisingly, the managers who have full decision-making authority also rely more on informal practices. In addition, input into major HRM decisions does not seem to influence managers' identification with the business, nor their commitment to staying with their current operations.

Given the wide range of HRM decision-making authority across the middle managers interviewed, how do agricultural middle managers keep the work flowing smoothly, overcome problems, and maintain amicable relationships with and between employees? Each practice is briefly described in the first column, and an example from the interviews is presented in the second column. These excerpts present a glimpse of the reality of the interviewees, as well as the richness of the data. Since there are many tasks inherent in management jobs it is difficult to identify the full range of possible HRM practices.

Further, many of the tasks tend to be points on a continuum, rather than discrete categories that can be readily distinguished one from the other. For example, a manager might communicate with an employee about a deadline for a particular task. However, this type of communication could also be described as feedback, since the manager might mention that this deadline is "more firm" than the deadline for a previous task that the employee had missed.

Finally, this communication could also be an opportunity for on-the-job training, as the manager might indicate how the employee's work speed can be improved. Thus, these categories are not meant to be exclusive, but instead present the full range of the practices the middle managers described in their interviews.

Practices have been classified as traditional based on their correspondence with traditional HRM functions as discussed in textbooks. However, these functions were adapted to specific agribusiness contexts, as well as to the middle management level, because the decision-making authority of many of these managers is rather limited For example, the labor relations function typically arises in dealing with conflict, not in formal union contracts. In contrast, managing relationships with employees is classified as a participative practice because the way it is used has little in common with traditional approaches to labor relations.

Personal Selling

The person who sells goods to you in this way is called a 'salesman' and the technique of selling is known as 'personal selling' or 'salesmanship'. Thus, personal selling refers to the presentation of goods before the potential buyers and persuading them to purchase it. It involves face-to-face interaction and physical verification of the goods to be purchased. The objective is not only just to sell the product to a person but also to make him/her a permanent customer.

Think about the salesman who come to you to sell goods and commodities. What do they do?

They show certain array of goods to you, try to explain the features of the products, if required demonstrate the functioning of the items, inform you about the price concession available, persuade you to buy the product and also in some cases promises you to bring certain items of your choice in future. So not only do they inform and explain to you about the product but also persuade you to buy those items and want you to buy from them in future also. On the other hand, you also gather more information about the product, see and handle it personally to judge it better.

You can also find personal selling in some shops where salesmen are employed by the shopkeeper to use this technique. For example, you can find such salesmen in jewellery stores, consumer goods stores, saree houses, etc. In case of some services, we also find personal selling used in shops. For example, we find people going to the same barbershop to cut their hair and get a massage from a specific barber. This shows that in case of personal selling the seller usually come to know about the taster and preferences of the customer and thus attracts him to buy the goods or services.

Essential elements of Personal Selling

Personal selling consists of the following elements:

i. Face-to-Face interaction: Personal selling involves a salesmen having face-to-face interaction with the prospective buyers.

ii. Persuasion: Personal selling requires persuasion on the part of the seller to the prospective customers to buy the product. So a salesman must have the ability to convince the customers so that an interest may be created in the mind of the customers to use that product.

iii. Flexibility: The approach of personal selling is always flexible. Sometimes salesman may explain the features and benefits of the product, sometimes give demonstration of the use of product and also faces number of queries from the customers. Looking into the situation and interest of the customers, the approach of the salesman is decided instantly.

iv. Promotion of sales: The ultimate objective of personal selling is to promote sales by convincing more and more customers to use the product.

v. Supply of Information: Personal selling provides various information to the customers regarding availability of the product, special features, uses and utility of the products. So it is an educative process.

vi. Mutual Benefit: It is a two-way process. Both seller and buyer derive benefit from it. While customers feel satisfied with the goods, the seller enjoys the profits.

Importance of Personal Selling

Personal Selling is extremely important as it helps in increasing sales. But there are other features as well which make it important. Let us discuss the importance of personal selling from the point of view of manufactures as well as consumers.

From manufacturer's point of view

i. It creates demand for products both new as well as existing ones.

ii. It creates new customers and, thus help in expanding the market for the product.

iii. It leads to product improvement. While selling personally the seller gets acquainted with the choice and demands of customers and makes suggestions accordingly to the manufacturer.

From customer's point of view

i. Personal selling provides an opportunity to the consumers to know about new products introduced in the market. Thus, it informs and educates the consumers about new products.

ii. It is because of personal selling that customers come to know about the use of new products in the market. The sellers demonstrate the product before the prospective buyers and explain the use and utility of the products.

iii. Personal selling also guides customers in selecting goods best suited to their requirements and tastes as it involves face-to-face communication.

iv. Personal selling gives an opportunity to the customers to put forward their complaints and difficulties in using the product and get the solution immediately.

A BUSINESS FOR MULTITALENTED PEOPLE

Like so many other sectors, agriculture is evolving fast. And naturally, so is the trade in farm produce. Not only is it characterised by rapid technological development, it also takes place on an increasingly global scale. In short, it is an ideal arena for top-level agribusiness managers - experts in buying and selling. People like you, perhaps.

Altogether, farming covers an enormous sector and accounts for a large flow of goods. From seeds to crops, from machinery to animal feeds, everything the sector needs or produces has to be bought, transported, and sold.

To work successfully in this sector, you will need a variety of personal qualities. In agribusiness, a wide range of situations can arise in a single day, and it takes talent to manage them. That is why, rather than focusing on traditional types of academic knowledge, the programme stresses professional competencies: problem identification, problem solving, communication skills, processing information, teamwork - everything you really need to make a difference.

Personal Selling

Occurs through personal communication in an exchange situation.

- More specific communication aimed at one or more persons.
- Effective at building buyers preferences, convictions and actions.
- Cost per person is high, most expensive promotional tool.
- Greater impact on consumers
- Provides immediate feedback
- Allows marketers to adjust message quickly to improve communication.
- Buyer feels a great need to listen and respond.
- Long term commitment is needed to develop a sales force.

AGRICULTURAL ECONOMICS AND AGRIBUSINESS

The agricultural business degree program provides education suited to career opportunities in farm management, agricultural business management, and agricultural marketing in both the domestic and international areas.

Managers of farms and agricultural businesses are continually required to make organizational and operational decisions. The basic skills and knowledge needed for making sound decisions are provided by the agricultural business curriculum. Students may elect to specialize in areas compatible with their personal objectives, depending upon the extent of accounting and business orientation desired.

Students educated in agricultural business are in demand for positions in agricultural industries, farm operation, marketing agencies, agricultural service organizations, state and federal agencies, and numerous other positions. For those who go on to graduate school, teaching and research positions are available with land grant colleges as well as with other institutions. Three concentrations are available to meet career objectives:

A. Agricultural Business Management and Marketing (ABMM)
B. Pre-Law, for students preparing to attend law school (PRLW)
C. Agricultural Economics, which emphasizes quantitative and analytical skills to prepare students for graduate school (AGEC).

Personal selling is one of the oldest forms of promotion. It involves the use of a sales force to support a push strategy (encouraging intermediaries to buy the product) or a pull strategy (where the role of the sales force may be limited to supporting retailers and providing after-sales service).

What are the main roles of the sales force?

Kotler describes six main activities of a sales force:

(1) **Prospecting** - trying to find new customers

(2) **Communicating** - with existing and potential customers about the product range

(3) **Selling** - contact with the customer, answering questions and trying to close the sale

(4) **Servicing** - providing support and service to the customer in the period up to delivery and also post-sale

(5) **Information gathering** - obtaining information about the market to feedback into the marketing planning process

(6) **Allocating** - in times of product shortage, the sales force may have the power to decide how available stocks are allocated

What are the advantages of using personal selling as a means of promotion?

- Personal selling is a face-to-face activity; customers therefore obtain a relatively high degree of personal attention
- The sales message can be customised to meet the needs of the customer
- The two-way nature of the sales process allows the sales team to respond directly and promptly to customer questions and concerns
- Personal selling is a good way of getting across large amounts of technical or other complex product information
- The face-to-face sales meeting gives the sales force chance to demonstrate the product
- Frequent meetings between sales force and customer provide an opportunity to build good long-term relationships

Given that there are many advantages to personal selling, why do more businesses not maintain a direct sales force?

Main disadvantages of using personal selling

The main disadvantage of personal selling is the cost of employing a sales force. Sales people are expensive. In addition to the basic pay package, a business needs to provide incentives to achieve sales (typically this is based on commission and/or bonus arrangements) and the equipment to make sales calls (car, travel and subsistence costs, mobile phone etc).

In addition, a sales person can only call on one customer at a time. This is not a cost-effective way of reaching a large audience.

Skills for successful selling

Interpersonal skills needed for selling

Now let´s identify the specific interpersonal skills and abilities you need so you can sell effectively. Here again are the five steps of selling, with the interpersonal skills appropriate for each stage indicated.

You can´t get past first base unless you can meet and greet people successfully. Additionally, the above diagram shows that you use certain key interpersonal skills in almost all the stages of selling.

If you click on the image, it opens a popup window displaying the image. Then you can move it to the top right-hand corner of your screen so it´s still there for reference as you move down this rather long page.

Meeting and greeting

Clients buy you before they buy your product or service. They need to feel comfortable with you so they´re comfortable discussing their needs and wants with you.

Clients judge you within ten seconds, basing their judgement on your appearance, communication and attitude.

This means, before you can even begin the five personal selling steps, you need to meet and greet appropriately. For example:

- Make eye contact.
- Give a warm smile.
- Give a sincere, friendly greeting.
- Pay attention to the client rather than to the product or something else.
- When appropriate, ask a useful opening question or make an appropriate statement. For example, ´Welcome to our store. What are you looking for today?´
- Never ignore the client.
- Best of all: look forward to meeting them and want to meet them.

Building rapport

Rapport has many definitions, the simplest probably being ´to be able to relate to and make a connection with another person´.

Clients buy from people they like and trust. So when you can make others feel comfortable with you by developing rapport with them, they´re more likely to want to buy from you.

As you build rapport people become more willing to share their needs with you. They may also come to trust your recommendations regarding the products and services you suggest to them. And you encourage repeat visits.

Questioning

To sell, you also need questioning skills. When you ask the right questions you can move through the personal selling steps quite fluidly. Asking appropriate questions has a very important byproduct for the client as well: they perceive that you're interested in them.

Listening

When you're selling, you identify your client's needs so you can suggest an appropriate product or service to meet those needs. This means you must listen to them.

Active listening is the process of confirming what you think the speaker meant as you heard their words and observed their verbal and non-verbal cues. It involves, in addition to other things, paraphrasing what you think the other person said. It requires:

- focusing your full attention on your client
- paraphrasing your understanding of their meaning
- taking notes if necessary
- using appropriate non-verbal cues such as nodding your head, inclining your body forward and maintaining eye contact
- noting their non-verbal cues
- using appropriate and well timed probing questions and summary confirmation questions.

Your listening skills are important to identifying your client's needs, making your sales presentation, closing your sale and offering after-sales services.

Non-verbal communication

Communication is 7% verbal, 38% the way you say things (verbal style) and 55% body language. This means that the person receiving your message gathers most of the message from something other than your actual words.

What you say isn't nearly as important as how you say it. Non-verbal communication can be split into two areas: body language and verbal style. Successful salespeople communicate well non-verbally themselves and they are able to pick up on their clients' non-verbals.

Body language includes

- facial expressions (e.g. wrinkling the nose, furrowing the brow or rolling the eyes)
- amount of eye contact
- smile
- hands (e.g. whether they're near the face)
- gestures (e.g. open or closed arms or nodding the head)
- posture (e.g. standing upright, slouching or inclining the body forward)
- position (e.g. observing personal space and accommodating cultural differences).

Verbal style includes

- pitch
- volume
- tone
- speed
- inflection and enunciation
- quality and intensity
- meaning conveyed by the sound of the voice.

Both those lists immediately above apply equally to both yourself as the salesperson and to your client.

When words and body language don´t match, the listener becomes confused and pays more attention to body language.

CONVERSATIONAL SKILLS

You need to be able to develop rapport, ask questions, identify needs, explain product benefits and overcome objections in order to move through the personal selling steps. These are conversational skills.

If you have good conversational skills, you:

- talk knowledgeably about your product or service
- display interest and warmth
- avoid bias or stereotyping
- adjust to the other´s verbal style
- tell the truth
- minimise your own opinions
- accept the other´s opinions
- don´t interrupt or correct unnecessarily
- watch for and respond to signs of discomfort or boredom
- are diplomatic
- can make small talk when it´s appropriate.

Ability to put yourself in the other person´s shoes.

This is what we shortened to ´understanding´ in the image above.

We all have certain expectations when we deal with a salesperson. We expect a salesperson to:

- Be able to meet our needs and solve our problems.
- Help facilitate an enjoyable experience.
- Be knowledgeable about the products and services they offer to us.
- Provide good client service to us.
- Be courteous to us.

When you are able to put yourself in another person′s shoes, you can think about what that person expects from you. When you take the time to think about the person′s expectations, you are better positioned to move through the selling steps in a way that meets those expectations.

SKILLS FOR SUCCESSFUL SELLING

Skills and characteristics of successful salespeople

Now we′ve covered successful selling skills and characteristics of successful salespeople. To a very large extent these skills and characteristics go hand-in-hand. When someone increases a selling skill, they begin to work on the characteristics that successful salespeople have. The process works the other way around as well.

For example, suppose Tina has a habit of twirling her hair when talking to clients. Some clients might perceive that she′s not serious about what she′s saying. Some might even think she′s insincere. Her lack of awareness of a particular non-verbal communication skill (body language) is impacting client perception of her characteristics.

When she corrects this, she automatically enhances her clients′ perception of her sincerity. Personal entails the face-to-face pitching of a product or service to a prospective buyer. The main thing that sets personal selling apart from other methods of commerce is the intensive interpersonal skills required, given that that the salesperson conducts his or her business with the customer in person.

Personal selling dates back to the Bronze Age. Traveling sales kits made up of bones and stones have actually been found from this era. In the United States, the first salesmen were Yankee peddlers who carried their goods from the east on their backs. They traded clothing, spices, pots and pans, and other household goods to settlers on the western frontier.

The father of modern selling techniques is considered to be John Henry Patterson, the head of National Cash Register company. As far back as the late 1800s he was implementing sales training programs, quotas, and sales territories for his sales staff. He also introduced the notion of canned sales talks.

Personal selling is one part of the promotion mix, the various ways businesses choose to reach or communicate with their customers. The main elements in a promotion mix are advertising, sales promotion, public relations, and personal selling. Advertising is any form of paid presentation or promotion that is not done face-to-face, such as television commercials. Sales promotion is the use of incentives, such as coupons, to entice a customer to buy a product or service.

Public relations is the act of building up the image of a company in the eyes of the community in the hopes of translating the feelings of goodwill into sales. An example of public relations is a company sponsoring a charity event. The final component of the promotion mix is personal selling, in which a demonstration or presentation to a potential customer is performed in person.

Settling on a promotion mix involves many factors. Businesses may choose to use any or all of the promotion mix tools and must decide how to allocate resources for each component. Some of the things organizations should consider when deciding on a promotion mix are the type of product or service sold, the unit value of the product or service, and the budget allotted for the promotion mix. Of all the industries involved in the promotion mix, the personal selling industry involves the most people. As a comparison, there are about 500,000 people involved in the advertising industry but more than 13 million people in personal selling.

In general, if a product has a high unit value and requires a demonstration, it is well suited for personal sales. For example, an encyclopedia is a high-priced item and most people do not feel they need one. After a demonstration, however, most people agree it would be a useful item to have.

Therefore, encyclopedias are well suited to a promotion mix that emphasizes personal selling. Highly technical products are also primarily sold through personal sales methods. Computers and copiers are good examples of technical products that are best sold through personal sales. Products that involve a trade-in are also best sold through personal selling to help facilitate the trade-in process.

Automobile sales often involve a trade-in and almost always involve a personal sales transaction. Finally, an organization that cannot afford an advertising campaign which is a very expensive endeavor might consider personal selling as an alternative to advertising. A personal sales force is a relatively inexpensive alternative to advertising, as the primary cost is the sales-force compensation. Since sales-force compensation is largely based on actual sales, a sales force is an investment that requires much less money up front than do other forms of the promotion mix that need time to pay off.

Personal selling as a career is unique and offers many benefits. It is, however, not for everyone. In general, it involves long, irregular work hours and extensive travel. A personal salesperson should also be able to handle rejection face to face, which is a large component of the job. On the other hand, personal sales offers great rewards for those who are successful. Because most compensation involves commissions based on completed sales, the potential for income is great. With personal sales, there is no ceiling on what a person can earn, as there is with other salaried jobs.

Also, many people enjoy the freedom of flexible hours and the fact that a personal salesperson has little contact with a supervisor. A career in personal sales offers a person the chance to develop interpersonal, communication, organizational, and time-management skills.

SALES-FORCE COMPENSATION

It varies from one organization and industry to another. All compensation plans, however, contain one or more of the following components: commission, bonuses, expense accounts, incentives, benefits, and a salary or draw. Commissions, by which a salesperson is paid a percentage of the sale he or she makes, is the most common type of sales force compensation because it directly ties compensation to performance. Bonuses based on performance are often employed as well. With expense accounts or allowances, some companies will reimburse salespeople for business expenses incurred.

Another form of compensation, and one that can be extremely motivating for some people, is incentive prizes earned through sales contests. Cars, trips, cash, and a number of other prizes are offered in exchange for meeting certain sales goals. Many companies offer benefits such as life and health insurance, although these benefits too can be tied to sales performance. Finally, some companies pay a base salary or draw, usually in conjunction with one or some of the other compensation elements. A draw is a fixed amount that is held against a salesperson's future sales earnings. This is usually offered to a new salesperson in order to foster earnings stability while he or she is learning the business. Usually, if the salesperson does not make future sales, he or she is not held responsible for the amount.

In general, the tighter the control a company has over a salesperson, the larger the role salary plays in compensation. For example, an IBM salesperson, based in a branch office and receiving extensive training and supervision, may have a large part of his beginning compensation plan made up of a base salary or draw. At the other end of the spectrum you may find a World Book salesperson, based in her home, with little training and supervision. She may never even see a branch office and will be compensated entirely in commissions and bonuses.

Types of Personal Sales Jobs

Noted industrial psychologist Robert McMurry identified the main types of personal sales jobs:

- *Driver-sales person*: This person merely delivers the product and has few selling responsibilities.
- *Inside order taker*: In this position, a person takes orders from within a selling environment. Examples include a sales clerk in a retail store, or a phone representative working for a catalog sales company. Some selling skills are required.
- *Outside order taker*: These salespeople go to the customer's place of business and take orders. Most of these sales are repeat business. Some selling is required, especially to establish new accounts.
- *Missionary sales person*: This type of sales involves selling goodwill but not any actual product or service. This salesperson's goal is to make a customer feel good about the company, products, or services the salesperson represents. Companies in the pharmaceutical and liquor industries, for example, typically employ missionary salespeople.
- *Sales engineer*: These positions are found in technical industries such as computers and copiers. Sales engineers provide technical support, explain the products, and adapt the product to the customer's needs.

- *Creative salesperson*: These salespeople attempt to sell goods (vacuum cleaners or encyclopedias), services (insurance), or causes (charities). These salespeople are usually dealing with customers who are unaware of their need for the service or product, and so, the salespeople must possess the most refined selling skills.

Although, there are many different types of salespeople, they all go through the same basic steps when making a sale: prospecting and qualifying, pre-approach, approach, presentation and demonstration, handling objectives, closing, and follow-up. Although training criteria for personal sales forces may vary from one organization to another, most will include some version of these steps.

Prospecting and qualifying involve finding potential customers and determining whether they are in a buying position. Prospecting, or lead generation, can be as simple as asking current customers for names of acquaintances, or as sophisticated as using a database or mailing list. Often, the salesperson's company provides leads, but a truly successful salesperson will also be able to generate his or her own leads. Generally, prospecting involves an element of cold-calling—calling an unknown potential customer and introducing yourself and your product. Often, this is the least favorite part of a salesperson's job but ultimately one of the most important. In addition to prospecting for clients, a salesperson needs to qualify the customer. Is the potential customer financially in a decision-making position? Does this customer need the product or service? It is not atypical that a salesperson may need to contact many, many prospects before making a sale.

The pre-approach is the step salespeople take when they are researching their prospective customer—often another company. They may read up on the company, talk to other vendors, or find out more about the industry. The salesperson will also take time to set sales-call objectives and try to determine the best time to call.

The next step is the approach, which is crucial for a salesperson to start out on the right foot. The salesperson should introduce himself or herself, the company represented, and the product or service being offered. It is also important that he or she listen carefully to the prospect and respond appropriately.

Once the approach has been made the salesperson should be ready to launch into the demonstration or presentation. Depending on the company and the product or service, there are generally three types of presentations. The canned approach is a tightly scripted talk that is either memorized or read. The formula approach is less rigid and, depending on the buyer's response to some carefully asked questions, will be tailored to meet the customer's needs. The third presentation style is the need-satisfaction approach, in which the seller tries to find out the customer's needs mostly by listening.

Presentations and demonstrations may involve any number of visual aids, such as flip-charts or demonstrations of the products themselves. One of the keys to a successful presentation is product knowledge. The more the salesperson knows about the product or service, the more relaxed he or she will be, and the more able to answer questions, fill the customer's need, and handle objections.

Handling objections is the next phase of selling. Almost every customer will present objections to making a purchase, whether real or not. A good salesperson is not flustered by these objections and handles them in a positive, confident manner. One approach to objections, used frequently with canned presentations, is to simply acknowledge the objection and continue with the presentation. In the more tailored presentations, the salesperson can handle the objection by turning them into reasons to buy.

The next step of a sale is often identified by novice salespeople as the toughest step: closing, or asking the buyer to purchase. Some new salespeople are so reluctant to appear aggressive that they never try to close, and the customer may become annoyed and decide not to purchase for just that reason. Customers must be given the chance to purchase. Salespeople need to learn to look for signals that a closing is appropriate. Common signals that customers give include asking questions, making comments, leaning forward or nodding, and asking about price or terms.

The last step of a sale, the follow-up, is often neglected but is important for many reasons. The follow-up can be done in person or by telephone. This gives the customer the chance to ask questions and reinforce his or her buying decision. The salesperson can review how to use the product, go over instructions and payment arrangements, and make sure the product has arrived in proper working order. This step encourages repeat business, is a good opportunity to get referrals, and increases the chances that subsequent payments will be made.

Bibliography

Andrew H. Cobb; John P. H. Reade (2011). 7.1. Herbicides and Plant Physiology. John Wiley & Sons. ISBN 9781444322491.

EPA. February 2011 Pesticides Industry. Sales and Usage 2006 and 2007: Market Estimates Archived 2015-03-18 at the Wayback Machine.. Summary in press release here Main page for EPA reports on pesticide use is here.

Krämer, Wolfgang, ed. (2012). Modern crop protection compounds (2nd, rev. and enl. ed.). Weinheim: Wiley-VCH-Verl. pp. 197-276. ISBN 978-3-527-32965-6.

Lock, E. A.; Ellis, M. K.; Gaskin, P; Robinson, M; Auton, T. R.; Provan, W. M.; Smith, L. L.; Prisbylla, M. P.; Mutter, L. C.; Lee, D. L. (1998). From toxicological problem to therapeutic use: The discovery of the mode of action of 2-(2-nitro-4-trifluoromethylbenzoyl)-1,3-cyclohexanedione (NTBC), its toxicology and development as a drug. Journal of Inherited Metabolic Disease. 21 (5): 498-506. doi:10.1023/A:1005458703363. PMID 9728330.

Mallory-Smith, C. (1999). Impact of labeling herbicides by site of action: A University view. Weed Technology. 13: 662

Moran GR (Jan 2005). 4-Hydroxyphenylpyruvate dioxygenase (PDF). Arch Biochem Biophys. 433 (1): 117-28. doi:10.1016/j.abb.2004.08.015. PMID 15581571. Archived from the original (PDF) on 2014-03-03.

Protecting Crop Yields And Quality Worldwide.

Quastel, J. H. (1950). 2,4-Dichlorophenoxyacetic Acid (2,4-D) as a Selective Herbicide. Agricultural Control Chemicals. Advances in Chemistry. 1. pp. 244-249. doi:10.1021/ba-1950-0001.ch045. ISBN 978-0-8412-2442-1.

Retzinger Jr, E. J.; Mallory-Smith, C. (1997). Classification of herbicides by site of action for weed resistance management strategies. Weed Technology. 11: 384-393.

Robbins, Paul (2007-08-27). Encyclopedia of environment and society. Robbins, Paul, 1967-, Sage Publications. Thousand Oaks. p. 862. ISBN 9781452265582. OCLC 228071686.

Robert L Zimdahl (2007). A History of Weed Science in the United States. Elsevier. ISBN 9780123815026.

Schmidt, R. R. (1997). HRAC classification of herbicides according to mode of action. 1997 Brighton crop protection conference: weeds. Proceedings of an international conference, Brighton, UK, 17-20 November 1997, British Crop Protection Council. pp. 1133-1140.

Shaner, D. L.; Leonard, P. (2001). Regulatory aspects of resistance management for herbicides and other crop protection products. In Powles, S. B.; Shaner, D. L. Herbicide Resistance and World Grains. CRC Press, Boca Raton, FL. pp. 279-294. ISBN 9781420039085.

Stryer, Lubert (1995). Biochemistry, 4th Edition. W.H. Freeman and Company. p. 670. ISBN 978-0-7167-2009-6.

Van Almsick, A. (2009). New HPPD-Inhibitors - A Proven Mode of Action as a New Hope to Solve Current Weed Problems. Outlooks on Pest Management. 20: 27-30. doi:10.1564/20feb09.

Vats, S. (2015). Herbicides: history, classification and genetic manipulation of plants for herbicide resistance. In Lichtfouse, E. Sustainable Agriculture Reviews 15. Springer International Publishing. pp. 153-192.

Weed Science Society of America.

Zhou Q, Liu W, Zhang Y, Liu KK (Oct 2007). Action mechanisms of acetolactate synthase-inhibiting herbicides. Pesticide Biochemistry and Physiology. 89 (2): 89-96. doi:10.1016/j.pestbp.2007.04.004.

Index

A

Actors 11, 32
Agri-food 10, 13, 116
Agronomics 25
Analysis 61, 198, 207, 214, 217, 219
Apples 50
Assets 13, 108, 169, 200, 202, 207-208
Assignment 225

B

Behavior 128, 239
Beverages 145
Biodiesel 59
Bioethanol 59
Biology 14
Biomass 59
Boards 57, 82
Brokers 91
Budget 155-156, 158, 204, 220
Business 3-5, 19-21, 23, 38, 81, 85, 110, 122, 124-125, 155, 161, 165, 167-173, 184, 207-208, 210-212, 214, 216-220, 224, 228, 236, 244, 251

C

Capacity 95, 104
Capital 36, 43, 134, 177, 194, 196, 198-206, 230
Carbon 68
Carcass 101
Cash-flow 220
Castes 233
Category 127
Census 204
Century 25, 30, 64
Cereals 102
Chickens 66
Climate 18
Commerce 6
Commercial 56, 84
Committee 91
Commodity 2, 50, 99, 102, 105, 174, 202
Community 39-40, 167
Components 128
Computer 7, 209
Consumers 2-3, 114, 120, 145, 243
Culture 234, 236
Customers 76, 242-243, 245, 252

D

Databases 215
Declines 139
Delphi 126
Demands 113, 116

Density 151

Designs 214

Dimensions 234

Disease 114

Dividends 209

Doctor 46

Documents 207

E

Ecological 182

Economic 34-36, 38-40, 182, 189, 200

Economics 25-26

Economy 40-43, 182, 201

Ecosystem 71

Education 205

Electronic 212

Employees 24, 164, 224, 228-230, 232, 241

Employment 43, 236

Energy 58-60, 139

Entity 165

Equipment 160, 203

Equity 208

Experts 126

Exports 44

Extension 28-34, 63, 223

F

Factors 4-5, 49, 53

Family 27, 175

Farm-based 176

Farmers 8-9, 12, 30-32, 34, 41-42, 56-57, 62-63, 65, 78, 80, 84-85, 87-88, 90-92, 94-95, 98-100, 102, 111, 123, 156, 163-164, 180-181, 223

Farmland 115

Fertilizer 206

Finance 62, 218

Fishing 204

Foodgrains 44

Forecasts 125-126

Framework 118

Future 48, 121, 124-125, 127, 135, 198

G

Gasoline 126

Global 16

Governance 20

Government 8, 42, 89, 91-93, 96, 102, 105, 107, 109-111

Grades 104-105

Grading 91, 93, 98, 100, 105

Grains 100, 102

Grazing 67, 71

Growing 37, 59, 81, 122

Growth 3, 35-38, 56, 193, 204

H

Harvest 20, 34

Health 65, 67, 181

Hemisphere 118

Herbicides 255-256

Heritage 233

High-tech 185, 189

History 180

Holdings 41, 142-143, 149-150, 152-153

Households 140-141, 201

I

Indicators 140

Indices 134

Industry 3, 5, 34, 43, 109, 122, 182-183, 185-186, 190, 201, 250

Informal 241
Insurance 158
Integrated 72
Investment 74-76, 118, 166, 196-197
Irrigation 69

K

Kingdom 76, 133, 142, 149
Labour 47, 136, 139, 151, 235

L

Leaders 224
Liability 169, 171
Livestock 66, 71, 100-101, 151, 203-204
Logistics 159-162

M

Machinery 203, 206
Management 4, 21-22, 24, 67, 78, 122, 162-163, 191-192, 209-210, 212-213, 216-220, 224-232, 235-241, 244
Mandies 90
Market 4, 11, 52, 55-56, 76, 78-80, 82-84, 87, 91-93, 95, 97-98, 103, 105, 108-109, 173, 183, 186, 243
Markets 11, 17, 55-56, 58, 78-81, 83, 85, 87, 90-93, 99, 109, 117, 119, 184, 186
Maturity 193
Medical 189
Models 25, 194
Module 231-232
Movement 181
Movements 159

N

Nitrogen 68
Non-verbal 247
Nutrition 157, 159

O

Offers 250
Operations 7, 66, 162, 183, 215, 238
Organic 71

P

Paddies 70
Parastatal 102
Partners 171-172
Pastures 68
Pathways 12
Payments 158, 171
Payroll 231
Personnel 224, 226, 229
Pesticides 178, 206
Phases 213
Planning 4, 75-76, 93, 124-125, 164, 213, 219
Plants 58
Policy 11, 157
Population 41, 43, 45, 49, 176
Poultry 101
Principles 120
Processors 114, 177
Products 2-3, 44, 46, 79, 88, 98-99, 101, 119, 132, 134, 143-145, 166, 185, 187-188, 192-193, 195-196, 243, 250
Promotion 249-250
Protection 255

Q

Quality 91, 116, 119, 160-161, 186
Quantity 194

R

Rapport 246
Records 209-211, 220, 231

Registry 77
Regulation 147
Relocation 228
Repayment 220
Research 4, 10, 40, 93, 189, 205, 237-238, 240
Resistance 256
Returns 169
Revenues 155
Revolution 178-179
Rotation 69

S

Safety 113-114, 160, 192, 194-195, 236
Salesman 242
Satisfy 46
Sectors 189, 200-201, 206
Security 15, 215
Seller 243
Senders 32
Skills 244, 246, 248-249
Software 209, 215, 230
Soybeans 166
Standards 105, 216
Statistics 147-148
Supply 8, 61, 107, 191, 196
Surveys 91, 147
Systems 10, 13, 18, 27, 31, 67-68, 71, 100, 102, 114, 119-120, 162, 211-213, 215-216, 230-231

T

Taxation 169
Theory 240
Tobacco 145
Tourism 183
Traders 56, 83-84
Training 223, 229-230, 232
Transgenic 179

V

Values 134, 140, 206, 234
Variables 127, 235
Verbal 247
Visitors 164

W

Warehouses 96
Wealth 37
Wholesale 56, 83, 90
Workforce 142, 235